T0234315

Frontiers in Applied Dynamical Systems: Reviews and Tutorials

Volume 9

Frontiers in Applied Dynamical Systems: Reviews and Tutorials

The Frontiers in Applied Dynamical Systems (FIADS) covers emerging topics and significant developments in the field of applied dynamical systems. It is a collection of invited review articles by leading researchers in dynamical systems, their applications and related areas. Contributions in this series should be seen as a portal for a broad audience of researchers in dynamical systems at all levels and can serve as advanced teaching aids for graduate students. Each contribution provides an informal outline of a specific area, an interesting application, a recent technique, or a "how-to" for analytical methods and for computational algorithms, and a list of key references. All articles will be refereed.

Jean-Luc Thiffeault

Braids and Dynamics

 Springer

Jean-Luc Thiffeault
Department of Mathematics
University of Wisconsin Madison
Madison, WI, USA

ISSN 2364-4532 ISSN 2364-4931 (electronic)
Frontiers in Applied Dynamical Systems: Reviews and Tutorials
ISBN 978-3-031-04789-3 ISBN 978-3-031-04790-9 (eBook)
https://doi.org/10.1007/978-3-031-04790-9

Mathematics Subject Classification: 37-02, 37A35, 37B40, 37E30, 20F36

This Springer imprint is published by the registered company Springer Nature Switzerland AG
The registered company address is: Gewerbestrasse 11, 6330 Cham, Switzerland

Dedicated to the memory of Hassan Aref

Preface

I first learned about the applications of topology to fluid mixing in the paper by Boyland, Aref, and Stremler [28]. Their simple experiment, which will be described in Chap. 1, was a powerful illustration that the global features of rod motion could have an important effect on mixing, and that sometimes two motions that looked superficially similar could have dramatically different properties. What struck me most was how intuitive these ideas were—it was easy to visualize everything, even if it wasn't always clear how to go about computing quantities of interest.

I have spent many of the following years getting familiar with William Thurston's beautiful ideas. As a physicist turned applied mathematician, this didn't come naturally for me. The pure mathematics literature is difficult to approach from any angle, especially if one wasn't raised in it. In this short book, I've tried to distill the principal ideas that I think are essential in understanding topological dynamics, or at least that are most important to applications. The applications I have in mind are mostly to mixing in fluid dynamics, and to the more esoteric area of taffy pullers. Taffy pullers are not exactly a big business these days, but they are such visually striking representations of pseudo-Anosov maps that they are worth studying on their own merits.

The book does not make much of an explicit connection between topological ideas and fluid dynamics, in the sense that there are no differential equations herein. This connection is hard to make in practice when it comes to fluid mixing, and there are still many open questions. I have made the topological part more prominent, to keep the book more accessible and focused. In addition, more current applications involving braids from data are discussed only briefly in the last chapter, since the theory for dealing with these statistical or random braids is not yet well developed.

This book is aimed at beginning graduate students, advanced undergraduates, or practicing applied mathematicians looking to learn more about the topic, to serve as a starting point for more advanced ideas. It will also be useful for pure mathematicians in search of a rich source of motivating examples. Some basic knowledge of algebraic topology is useful, in particular homotopy and homology groups, but the essential concepts are reviewed briefly.

There are a number of people I'd like to thank, either for their direct help or for their role in the development of the ideas in this book. Phil Boyland, Toby Hall, and Erwan Lanneau have been invaluable over the years in answering my questions as I grappled to understand the material—with the usual disclaimer that the blame for inaccuracies rests solely with me. Matt Finn and Emmanuelle Gouillart were my constant companions during work on this subject in the early days, and I miss collaborating with them tremendously. Mark Stremler has always been full of insight on where to go next with all this. Jacques-Olivier Moussafir introduced me to Dynnikov coordinates after he unearthed them on a trip to Russia, and proposed the method for computing topological entropy that we use today. More recently, Marko Budišić has helped guide my thinking and has been my co-conspirator in creating the software package braidlab. Spencer Smith has let me use his nice notes on Dynnikov update rules as an appendix in the book. Many others deserve mention for stimulating discussions or for commenting on this book: Michael Allshouse, Persi Diaconis, Margaux Filippi, John Lynch, Kevin Mitchell, Tom Peacock, Sarah Tumasz, and Oyku Yurttaş. Ken Ribet kindly provided me with his photo of Thurston and Sullivan's mural at Berkeley. I'm also grateful to Chris Jones and Achi Dosanj for suggesting I write this book, at the 2013 SIAM Snowbird meeting. It took me a few years to get there, despite the short length. Finally, some of the research for the material in this book was supported by the US National Science Foundation, under grant CMMI-1233935.

Madison, WI, USA   Jean-Luc Thiffeault
November 2021

Contents

List of Symbols

Surfaces

S	An orientable surface
$S_{g,n}^b$	Orientable surface of genus g with n punctures and b boundaries (omit b or n if zero)
T^2	Two-dimensional torus
D_n	Disk with n punctures
∂D	Boundary of disk D_n
$\chi(S)$	Euler characteristic of surface S
$\widetilde{S}$	Covering space of S

Diffeomorphisms

ϕ, ψ, χ	Diffeomorphisms
$\mathrm{Diff}(S)$	Group of diffeomorphisms of a surface S
$\mathrm{Diff}^+(S)$	Group of orientation-preserving diffeomorphisms of a surface S
$\mathrm{Diff}_0(S)$	Group of diffeomorphisms of a surface S isotopic to id
id	Identity map
ι	Hyperelliptic involution
$\mathrm{MCG}(S)$	$\mathrm{Diff}^+(S)/\mathrm{Diff}_0(S)$, the mapping class group of a surface S

Homotopy, Homology, and Curves

$\pi_1(S, x_0)$	First homotopy group (fundamental group) of S with basepoint x_0
H	Homotopy or isotopy
$H_1(S, \mathbb{Z})$	First homology group of S with coefficients in $\mathbb{Z}$
ϕ_*	Induced action of diffeo ϕ on $\pi_1(S)$
$L_{\mathscr{G}}(\cdot)$	Reduced word length in $\pi_1(S)$
$\mathrm{GR}(\phi_*)$	Growth in $\pi_1(S)$ under ϕ_*
α, β	Curves or simple closed curves, usually up to homotopy equivalence

$\lvert\alpha\rvert$	Length of curve α
e_i	Generating loop for $\pi_1(S, x_0)$
u_i	Free homotopy loops
$\{E_i\}$	Basis for $H_1(S, \mathbb{Z})$
$\tilde{e}_i, \tilde{E}_i, \tilde{x}_0$	Lifts

Braids

γ	Braid
B_n	Artin braid group with n strings
$\mathrm{br}(\phi)$	Braid representative for the diffeomorphism $\phi : D_n \to D_n$
σ_i	Standard braid group generators of B_n, $1 \le i < n$
Δ_n	Positive half-twist braid
$B(\gamma, t)$	Burau representation of braid γ with parameter t
$[\gamma](t)$	Reduced Burau representation of braid γ with parameter t

Matrices

I	Identity matrix
$\lvert M\rvert$	Determinant of matrix M
(m_{ij})	Matrix M with elements m_{ij}
$\lVert M\rVert$	Matrix norm $\max_i\left(\sum_j\lvert m_{ij}\rvert\right)$
$\mathrm{spr}(M)$	Spectral radius of matrix M
$\mathrm{SL}_2(\mathbb{Z})$	Special linear matrix group with integer coefficients
$\mathrm{PSL}_2(\mathbb{Z})$	$\mathrm{SL}_2(\mathbb{Z})/\{-I\}$, the projective version of $\mathrm{SL}_2(\mathbb{Z})$

Train Tracks

G	Train track graph
$a, b, c \ldots$	Oriented main edges of a train track
$1, 2, 3 \ldots$	Oriented peripheral edges of a train track
$\bar{a}, \bar{1}$	Inverse of oriented edges
g	Train track graph map
Dg	Derivative map of g
N	Transition matrix of main edges

Dynnikov Coordinates

$\mathscr{S}(S)$	Homotopy classes of essential simple closed curves on S
$\mathscr{S}'(S)$	Homotopy classes of essential simple closed multicurves on S
$\mathscr{D}_n(\mathbb{Z})$	Space of Dynnikov coordinates for $\mathscr{S}'(D_n)$
μ_i, ν_i	Intersection numbers with triangulation
u	Dynnikov coordinate vector for a multicurve
a_i, b_i	Components of $u = (a_1, \ldots, a_{n-2}, b_1, \ldots, b_{n-2})$
$\ell(u)$	Minimum length of a multicurve u

$L(u)$	Minimum number of intersections of a multicurve u with the real axis
f^+, f^-	$\max(f,0), \min(f,0)$
$\oplus, \otimes$	Max-plus algebra addition and multiplication
$\mathbb{O}, \mathbb{I}$	Max-plus algebra zero, unit
$[\![\cdot]\!]$	Interpret expression as max-plus (e.g., $[\![(x+y)/z]\!] = (x \oplus y) \otimes z^{-1}$)

Miscellaneous

$\times$	Direct product
$\ltimes$	Semidirect product
$\mathbb{Z}, \mathbb{Z}_{\geq 0}$	Integers, nonnegative integers
$\mathbb{R}$	Real numbers
$\mathbb{C}$	Complex numbers
$\mathbb{E}$	Euclidean plane
$\mathbb{Z}[t,t^{-1}]$	Ring of integer polynomials in t and t^{-1} (e.g., $3t^{-2}+2-6t^3$)
$\mathbb{I}$	Unit interval $[0,1]$
$\simeq$	Isotopy equivalence
$\approx$	Isomorphic; "approximately equal to" for real numbers
π	Projection map
G	Group (sometimes graph)
$\mathscr{G}$	Set of generators $\{e_1,\ldots,e_n\}$ for a Finitely generated group G
F_n	Free group over n symbols (e.g., $e_1 e_2^2 e_4^{-1} \in F_4$)
$\mathbb{Z}F_n$	Group ring with integer coefficients (e.g., $e_1 e_3^{-1} + 2e_2^2 e_1 - e_4 \in \mathbb{Z}F_4$)
$\mathrm{Aut}(F_n)$	Group of automorphisms of F_n
$\frac{\partial}{\partial e_j}$	Free derivative on $\mathbb{Z}F_n$
$\mathscr{F}_\mathrm{s}, \mathscr{F}_\mathrm{u}$	Stable, unstable foliations
$\mu_\mathrm{s}, \mu_\mathrm{u}$	Stable, unstable transverse measures
$\lambda(\phi)$	Dilatation of a pseudo-Anosov diffeomorphism ϕ
$h(\phi), h(\gamma)$	Topological entropy of diffeomorphism ϕ or braid γ

Chapter 1
Introduction

*Shootings of water threads down the slope of the huge green
stone— The white Eddy-rose that blossom'd up against the
Stream in the scollop, by fits & starts, obstinate in
resurrection—It **is the life** that we live.*

—*Samuel Coleridge*

1.1 Motivation: Fluid Mixing

The mixing of fluids is crucial to many modern industrial processes. Often the goal
of mixing is to blend different reactants to enhance a chemical reaction; other times
dyes are mixed into a fluid, or a fluid is mixed to make it thermally homogeneous.
Geophysical flows—the motion of the Earth's ocean, atmosphere, and interior—also
involve many processes where mixing is important, such as the global biochemical
cycle that recirculates nutrients in the ocean, or the exchange of carbon dioxide
between the atmosphere and ocean due to breaking ocean waves.

At one extreme, the flows are *turbulent*. For example, the milk in your coffee
mixes rapidly because it only takes a flick of the wrist to make the flow turbulent.
Most of the mixing is due to vortices that swirl the milk and lead to rapid filamen-
tation. Turbulent mixing is also the dominant process for combustion in engines,
where many chemicals are caused to react quickly by mixing.

At the other extreme, flows are *laminar*, and these will be the focus of this book.
Laminar is the opposite of turbulent: the flow remains free of small-scale structures
such as tiny vortices, though aspects of the flow may still change in time. Laminar
mixing deals with very viscous fluids or fluids with complex properties, such as
polymers. Our natural intuition is that the mixing of viscous substances requires a
more careful approach. For example, if you are blending the ingredients for a cake,
do you drag the spoon as you would for a coffee cup, or do you instinctively make
a more deliberate motion, possibly following a characteristic "figure-eight" motion
with your spoon?

A controlled mixing experiment with a rod following a figure-eight path is shown
in Fig. 1.1. The fluid here is viscous corn syrup, and the dye is ink. The container
is circular and shallow, so the fluid motion can be idealized as two-dimensional.
A striking pattern of dye emerges, with the dye being "layered" by each repeated
rod motion. The reason this motion ensures good mixing is that as the streaks of
dye become thinner, molecular diffusion becomes more effective. With time, the

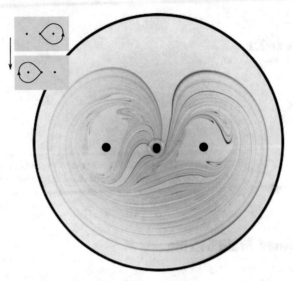

Fig. 1.1 The pattern made by stirring black dye into a clear viscous fluid, using a "figure-eight" rod motion. The inset indicates the motion of the middle rod around the other two rods (From Thiffeault et al. [122]; experiments by E. Gouillart and O. Dauchot.)

dark streaks will blend into each other and merge to yield gray fluid. Rod stirring accelerates this process tremendously.[1]

The laminar process involved with the figure-eight motion is very different from its turbulent counterpart. When turbulence is involved (i.e., when the rod moves fast or the fluid is less viscous), the rod sheds eddies, and it is these eddies that end up effecting the bulk of the mixing. The details of the rod motion do not matter much. By contrast, when turbulence is absent, there are no shed eddies, so the specific details of the rod motion matter greatly. This was dramatically demonstrated by Boyland et al. [27], whose experiment is shown in Fig. 1.2. Their experiment has three rods, as in the figure-eight experiment of Fig. 1.1. Observe that the dye in the left frame of Fig. 1.2 is much better mixed than that in the right frame. The two frames involve almost the same rod motion, except for a change in direction, indicated by the insets at the top left of each figure. We now give a simplified mathematical description that explains the essential difference between the two frames in Fig. 1.2. (We shall revisit this example in Sect. 3.4.)

Denote by $\overline{T}_1$ the clockwise interchange of the first and second rods, following a circular path, and by $\overline{T}_2$ the clockwise interchange of the second and third rods. The operation $\overline{T}_i^{-1}$ is the corresponding counterclockwise interchange. The motion $\overline{T}_2^{-1}\overline{T}_1$ corresponds to the rod motion in Fig. 1.4. (We read this expression

---

[1] For more details on viscous mixing, see the original chaotic advection article by Aref [4], the book by Ottino [96], or the recent review by Aref et al. [5]. The implications of stretching and folding to magnetic fields and dynamo theory are explored in the book by Childress and Gilbert [36].

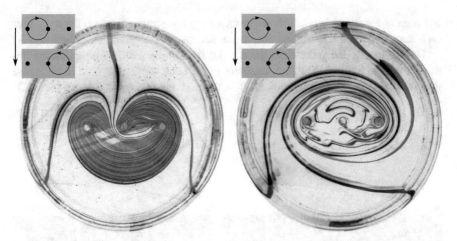

Fig. 1.2 Experimental stirring device with three rods immersed in a viscous fluid. Left: motion corresponding to $\overline{T}_2^{-1}\overline{T}_1$. Right: motion corresponding to $\overline{T}_2\overline{T}_1$. The insets show the direction of periodic rod interchange, with the arrow denoting time. The black dye suggests how the fluid is displaced (Adapted from Boyland et al. [27].)

from right to left since we will assign matrices to the $\overline{T}_i$ below, and matrices act on to their right.) Figure 1.2 (left) shows the effect of repeating this rod motion on three streaks of dye. The dye becomes layered, in a pattern almost exactly as in Fig. 1.1.

If however we instead use the rod motion $\overline{T}_2\overline{T}_1$, so that the rods are always interchanged clockwise, we obtain the pattern on the right of Fig. 1.2. No real layering occurs, and the mixing is very slow. The simple act of reversing a portion of the rod motion has turned this efficient mixing action into a terrible one. In fact, as we will see in Chap. 3, in this simple device, we can easily predict which motions are "good" by writing

$$\overline{T}_1 = \begin{pmatrix} 1 & 0 \\ 1 & 1 \end{pmatrix}, \qquad \overline{T}_2 = \begin{pmatrix} 1 & -1 \\ 0 & 1 \end{pmatrix}, \tag{1.1}$$

and then carrying out the matrix multiplication:

$$\overline{T}_2^{-1}\overline{T}_1 = \begin{pmatrix} 2 & 1 \\ 1 & 1 \end{pmatrix}, \qquad \overline{T}_2\overline{T}_1 = \begin{pmatrix} 0 & -1 \\ 1 & 1 \end{pmatrix}. \tag{1.2}$$

Both of these matrices have unit determinant. The first matrix in (1.2) is hyperbolic: its largest eigenvalue is $\frac{1}{2}(3+\sqrt{5}) > 1$. Hence, under repeated iteration, the spectral radius of the matrix $(\overline{T}_2^{-1}\overline{T}_1)^k$ grows exponentially with k, which is associated with the exponentially rapid layering observed in the experiment. By contrast, the matrix $\overline{T}_2\overline{T}_1$ in (1.2) has eigenvalues on the unit circle, so its spectral radius remains equal to one even under iteration, leading to very little layering in the mixing pattern. The matrix $\begin{pmatrix} 2 & 1 \\ 1 & 1 \end{pmatrix}$ is associated with the famous Arnold's Cat Map [6], which is a map of a torus to itself. This suggests a deep connection between the dynamics on a torus and three-rod stirring devices, which we will develop in Chaps. 2 and 3.

Unfortunately, for more than three rods, the simple matrix multiplication approach fails in general. The later chapters in the book will be devoted to this more general problem, where more powerful—and more interesting—tools are required.

1.2 Taffy Pullers

Fig. 1.3 (**a**) A taffy pulling device (from Finn and Thiffeault [50]). (**b**) Snapshots over a full period of operation, with taffy (from Finn and Thiffeault [50])

Fluid mixing is not the only application involving rod motion. A type of chewy candy called *taffy* is popular at seaside resorts in the USA. After it is made by heating sugar, taffy must be *pulled* repeatedly. Pulling is simply a process of stretching and folding, in the same way a baker kneads dough. The goal of pulling is to get air bubbles into the taffy, which gives it a nicer, chewier texture. Taffy pulling used to be laboriously done by hand, but in the early twentieth century many devices were invented to mechanize the process. Such a device is shown in Fig. 1.3a, and its operation is depicted in Fig. 1.3b.

Clearly, the operation of the taffy puller closely parallels the rod stirring devices we introduced in the previous section. The mixing pattern is replaced by the taffy itself. The rate of layering is the quantity we wish to find, or, equivalently, the rate of growth of the length of taffy. Because of the close analogy between taffy pullers and rod stirring devices, in this book we shall often not distinguish between them.

Our point of view is that we are after general mathematical principles and not after specific engineering designs. (Though in practice the latter issue is far from trivial!)

The history of taffy pullers is actually quite rich: the number and variety of patents that were filed proposing taffy pullers are impressive.[2] The main reason is that taffy used to be a big industry—one patent was acquired around 1900 for $75,000 (about two million of today's dollars). All these conflicting patents led to a cavalcade of lawsuits, which culminated in a 1921 ruling of the US Supreme Court, delivered by Chief Justice William Howard Taft. The opinion compares the merits of an older two-rod design by Firchau [54], and a newer three-pin device by Dickinson [40]:

> The machine shown in the Firchau patent [has two pins that] pass each other twice during each revolution [...] and move in concentric circles, but do not have the relative in-and-out motion or Figure 8 movement of the Dickinson machine. With only two hooks there could be no lapping of the candy, because there was no third pin to re-engage the candy while it was held between the other two pins. The movement of the two pins in concentric circles might stretch it somewhat and stir it, but it would not pull it in the sense of the art.

The Supreme Court opinion displays the fundamental insight that at least three rods are required to produce some sort of rapid growth. Moreover, the "Figure 8" motion is correctly identified as key to this growth.

As an illustration of the exhaustiveness displayed by the taffy patents, consider the simplest possible motion for a taffy puller, shown in Fig. 1.4a. This is the motion of the Boyland et al. [27] fluid mixing experiment of Fig. 1.2, as described in Sect. 1.1. The motion of the individual rods is shown in Fig. 1.4b (top). Notice that each rod follows a figure-eight motion. Though this is mathematically sensible, it is not very practical from an engineering standpoint: how can rods be made to move along two different circles?

In fact, a device that accomplishes exactly this figure-eight motion was invented by Nitz [94] in 1918 (Fig. 1.4b, middle). The device has two counter-rotating wheels, each with three slots for a rod. The three rods are made to alternate between the wheels by a tripping mechanism. It is not clear what advantages this design would have over the more standard design of Fig. 1.3a, but it seems important enough that it was reinvented again by Kirsch [73] in 1928. Such duplicate patents are not uncommon: there are at least five variants of the popular four-rod design shown in Fig. 1.5.

Of course, taffy pullers are rarely seen these days, and there is no longer a real industry built around them. However, there are analogous applications in the food industry with a much broader scope. One of these is the *mixograph*, shown in Fig. 1.6a. The mixograph consists of a small cylindrical container with three fixed vertical pins set in its base. A lid is lowered onto the base; the lid has four moving rods on compound gears, resulting in a net motion as in Fig. 1.6b. The mixograph is used to measure properties of bread dough. A piece of dough is placed in the device and is stretched by the motion of the pins. The torque on the rods is recorded on graph

---

[2] Many of the examples presented in this introduction are from my article *The Mathematics of Taffy Pullers* [118], which contains many more examples and additional historical and mathematical background.

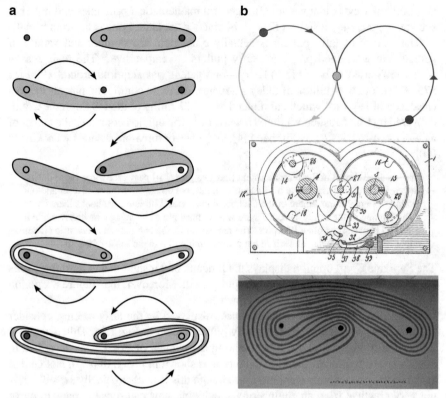

Fig. 1.4 (**a**) The action of a three-rod taffy puller (top to bottom). The first and second rods are interchanged clockwise, and then the second and third rods are interchanged counterclockwise. Two periods are depicted. (**b**) Top: Each of the three rods moves in a Figure-8. Middle: Taffy puller from the patent of Nitz [94]. Rods alternate between the two wheels. Bottom: The mural painted by Sullivan and Thurston at Berkeley in 1971 [87] (Photo by Kenneth Ribet.)

paper, in a manner reminiscent of a seismograph. An expert on bread dough can then deduce dough-mixing characteristics directly from the graph.

1.3 Outline

In this book, we will introduce the minimum amount of materials required to understand how rod motions give rise to exponential stretching, and how to compute the stretching rate. Some parts of this will be rigorous, and some will proceed by examples, though in general we avoid the definition-theorem-proof format in favor of a more discursive style. We will give references to sources with more complete proofs as we proceed. In particular, the book by Farb and Margalit [45] is invaluable for a student looking to study this subject more deeply. Some other great references

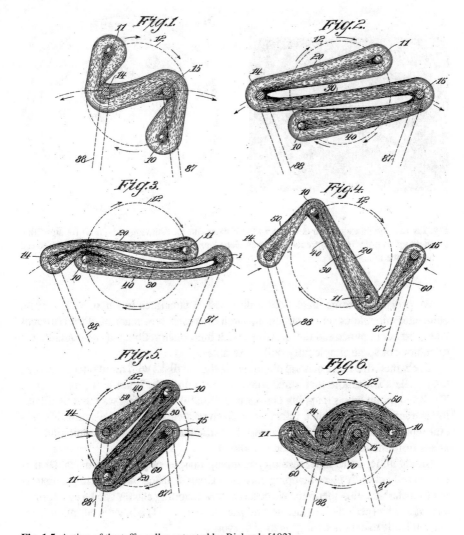

Fig. 1.5 Action of the taffy puller patented by Richards [102]

are the original book on braids and mapping class groups by Birman [18], the Orsay seminar notes by Fathi et al. [46] (translated to English by Kim and Margalit [47]), the short book by Casson and Bleiler [33], the reviews by Birman and Brendle [19] and Boyland [24], and the monograph on braids by Kassel and Turaev [72].

In Chap. 2, we will look at maps of the torus to itself. These serve as a prototype for what comes after, and everything can be computed explicitly. We introduce foundational concepts such as mapping classes and the mapping class group, as well as three fundamental topological classes of mappings: periodic, parabolic, and Anosov. We define the dilatation of an Anosov mapping class, a number that will play a decisive role throughout.

a   **b**

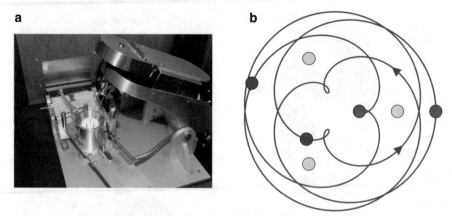

Fig. 1.6 (**a**) The mixograph, a device with three fixed and four moving rods, used for stretching bread dough in order to measure its properties. See Connelly and Valenti-Jordan [37]. (**b**) The rod motion for the mixograph

We project the torus maps onto a disk with 3 punctures in Chap. 3. This is an elementary double-cover construction, and it illustrates how maps of closed surfaces "descend" onto punctured disks. Armed with this construction, we can compute the dilatation for some simple taffy puller examples.

To characterize more general mapping classes on disks with more punctures (i.e., taffy pullers with more than three rods), we introduce the braid group in Chap. 4. The braid group labels mapping classes in a visually compelling way that facilitates the extraction of topological information from particle dynamics. We describe several representations of the braid group, in particular the Burau representation that arises from the action of braids on homology.

Braids allow us to label arbitrary mapping classes of punctured disks, so it is time to fully classify these mapping classes in Chap. 5, using the powerful Thurston–Nielsen classification theorem. We describe how mapping classes can be categorized as finite-order (periodic), reducible, and pseudo-Anosov. The latter is our main focus since it leads to exponential growth of curves.

Computing the growth of curves is now our central goal, and we relate this to topological entropy in Chap. 6. We discuss how topological entropy is bounded by growth on the fundamental group and how this estimate is exact for pseudo-Anosov maps. This gives us a way of estimating the dilatation from the Burau representation.

Better than an estimate is an exact calculation, and in Chap. 7, we introduce train tracks and show by examples how they are used to obtain exact results for the dilatation given as the largest root of a polynomial. We give a taste of the Bestvina–Handel algorithm for finding the train track corresponding to a given braid.

In Chap. 8, we present Dynnikov coordinates to describe equivalence classes of closed curves on punctured disks. The braid group acts on the coordinates by a simple piecewise-linear action, so they are well-suited to numerical computations. They are useful for deciding rapidly whether two braids are equal, and can also be used with iteration to get accurate estimates of the dilatation of a map. A companion appendix (by Spencer Smith) shows how to derive the update rules for the Dynnikov action of braids on closed curves.

Chapters 9 and 10 are devoted to introducing the Matlab library braidlab, which is used to manipulate braids and loops. In Chap. 9, we explain the basic functionality of the package, and in Chap. 10, we show how to create braids from two-dimensional orbit data. We discuss some of the difficulties that come with dealing with real-world data.

Chapter 2
Topological Dynamics on the Torus

Mmm... donuts.
—Homer Simpson

In this chapter, we use the torus to illustrate the basic ideas behind the topological classification of mappings. We introduce the mapping class group of the torus and investigate its properties. The torus is very special in that we can easily algebraically characterize all its mappings and give systematic examples. This will be a building block for 3-rod stirring devices (Chap. 3) and for more complicated cases in the remainder of the book.

2.1 Diffeomorphisms of the Torus

One of the simplest surfaces to study is the *torus*, denoted T^2. Its representation as a surface embedded in three dimensions is depicted in Fig. 2.1a, but for us it will often be more convenient to use the "flattened" view of Fig. 2.1b. The arrows in Fig. 2.1b indicate the edges that are identified with each other.

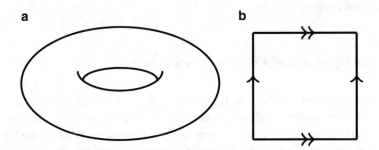

Fig. 2.1 (a) The humble torus. (b) A different view of the torus

In this chapter, the central question we want to answer is: what do maps of the torus to itself look like? The relevant maps are *homeomorphisms*: continuous, invertible maps whose inverse is also continuous. In fact, it costs us nothing to use instead *diffeomorphisms*: smooth, invertible maps whose inverse is also smooth. (See Farb

J.-L. Thiffeault, *Braids and Dynamics*, Frontiers in Applied Dynamical Systems:
Reviews and Tutorials 9, https://doi.org/10.1007/978-3-031-04790-9_2

and Margalit [45] for the reason why homeomorphisms can always be promoted to diffeomorphisms in the context of topology of surfaces.) The set of such maps forms a group, $\text{Diff}(T^2)$, where the group operation is simply composition of maps. We shall focus solely on $\text{Diff}^+(T^2)$, the group of orientation-preserving diffeomorphisms.

We will classify diffeomorphisms of a surface S up to *isotopy*, an equivalence relation denoted by $\simeq$. Two diffeomorphisms are isotopic if they can be continuously "deformed" into each other, such that the intermediate maps are also diffeomorphisms. More precisely, two diffeomorphisms ϕ and $\psi \in \text{Diff}^+(S)$ are *homotopic* if there exists a continuously varying family of functions H_t with

$$H_t : S \to S, \quad t \in \mathbb{I} = [0,1], \quad \text{with} \quad H_0 = \phi, \quad H_1 = \psi. \tag{2.1}$$

H_t is called a *homotopy*. If furthermore we have $H_t \in \text{Diff}^+(S)$ for all $t \in \mathbb{I}$, then H_t is an *isotopy* and $\phi \simeq \psi$.

Note that

$$\phi \simeq \psi \qquad \Longleftrightarrow \qquad \phi = \chi \circ \psi \quad \text{with} \quad \chi \simeq \text{id}, \tag{2.2}$$

that is, two maps ϕ and ψ are isotopic if and only if they are related by a map χ isotopic to the identity. Indeed, H_t in (2.1) gives an isotopy $H_t \circ \psi^{-1}$ from $\chi = \phi \circ \psi^{-1}$ to the identity; conversely, if H_t' is an isotopy from χ to the identity, then we can let $H_t = H_t' \circ \psi$ for an isotopy from ϕ to ψ. We write $\text{Diff}_0(S)$ for the subgroup of maps isotopic to the identity.

We define the *mapping class group* of a surface S by

$$\text{MCG}(S) := \text{Diff}^+(S)/\text{Diff}_0(S). \tag{2.3}$$

This is the central object that we wish to study: the mapping class group describes self-maps of a surface, up to isotopy. In that sense, it distills the essential topological information about such maps.

2.2 The Fundamental Group of a Surface

Diffeomorphisms of a surface can be hard to visualize. It is often much easier to instead look at the action of a diffeomorphism on representative curves on a surface S in order to deduce a map's main features. To that purpose, we briefly introduce two important algebraic objects: the fundamental group of a surface, and its associated first homology group. These topics are developed in more detail in textbooks such as Hatcher [69] and Munkres [89].

Let us start with a bit of terminology. A *curve* on a surface S is the image of a continuous map $\alpha : \mathbb{I} \to S$, where $\mathbb{I} = [0,1]$. The map α is called a *path*, and we typically use α to denote both the curve and its parametrized path. An *oriented* curve carries an orientation as well. A curve is *simple* if α is injective in $(0,1)$, so the curve does not intersect itself. A *closed* curve has $\alpha(0) = \alpha(1)$. An *essential* curve is not contractible to a point or to a boundary component of S. We shall often use the term *loop* for an oriented closed curve.

In the same way that homotopic maps were defined in the previous section, we can also define homotopic curves. Two curves defined by the paths $\alpha : \mathbb{I} \to S$ and $\beta : \mathbb{I} \to S$ are homotopic to each other if they share the same endpoints ($\alpha(0) = \beta(0)$ and $\alpha(1) = \beta(1)$), and there exists a continuously varying family of function H_t with

$$H_t : \mathbb{I} \to S, \quad t \in \mathbb{I} = [0,1], \quad \text{with} \quad H_0 = \alpha, \quad H_1 = \beta, \tag{2.4}$$

fixing the endpoints ($H_t(0) = \alpha(0) = \beta(0)$, $H_t(1) = \alpha(1) = \beta(1)$).

For a closed curve or loop, the "endpoint" is not well-defined, so we choose a distinguished point $x_0 \in S$, called the *basepoint*. Hence, a closed curve or loop α with basepoint x_0 has $\alpha(0) = \alpha(1) = x_0$. The definition of homotopic loops is then identical to homotopic curves above.

The *fundamental group* of a surface S with basepoint x_0, denoted $\pi_1(S, x_0)$, is the set of equivalence classes of loops with basepoint x_0, with the equivalence classes defined by homotopy of loops fixing the basepoint x_0. This is also known as the first homotopy group. For connected S, different choices of x_0 lead to isomorphic groups, so the basepoint x_0 is often dropped and we write $\pi_1(S)$.

The group operation in $\pi_1(S, x_0)$ arises because the loops are oriented. The composition of two loops is simply obtained by tracing one after the other, starting at x_0 and ending at x_0, possibly visiting x_0 several times in between. The inverse operation corresponds to reversing the orientation of a loop, that is, traversing its path in the opposite direction.

The fundamental group of S typically has an infinite number of elements since we can trace the same loop over and over again and always obtain a new loop. However, it is usually *finitely generated*, in the sense that all loops can be expressed as a product of a few loops and their inverse. Figure 2.2 shows a *generating set* for $\pi_!(T^2, x_0)$. These two loops have the property that they only visit x_0 at the start

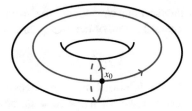

Fig. 2.2 Two nonhomotopic loops on the torus, with basepoint x_0. These form a generating set for $\pi_1(T^2, x_0)$

and end of their path. The generating set is of course not unique, though there is usually a simplest choice in terms of short loops.

The *first homology group* with coefficients in $\mathbb{Z}$ is denoted $H_1(S, \mathbb{Z})$. It is obtained from $\pi_1(S)$ by *Abelianizing*, that is, by assuming the group product is commutative. Thus, two loops α and $\beta \in \pi_1(S)$ with product $\alpha\beta$ correspond to two elements A

and $B \in H_1(S,\mathbb{Z})$ with "product" $A + B$. The group operation in $H_1(S,\mathbb{Z})$ is usually written as addition and subtraction to emphasize commutativity. As another example, the loop $\alpha\beta\alpha$ in $\pi_1(S)$ maps to the element $2A + B$ in $H_1(S,\mathbb{Z})$. The motivation for defining the first homology group is that it is a vector space, so that the power of matrices, eigenvalues, eigenvectors, etc. can be exploited. Its weakness is that the Abelianization process loses a lot of information about the group structure. We will see later (in particular in Chap. 4) how to use the homology group for practical calculations.

For the torus, the fundamental group $\pi_1(T^2)$ is Abelian, so the fundamental group and the first homology group coincide. This is not the case for more complicated surfaces.

2.3 The Mapping Class Group of the Torus

Having defined the mapping class group in Sect. 2.1, and the fundamental group in Sect. 2.2, we now visualize a diffeomorphism of the torus by its action on curves. This simply means that every point of a curve is mapped to some new curve on the same torus. Figure 2.3 shows the action of a simple diffeomorphism on two torus loops that intersect only once at their basepoint x_0. Note that the number of

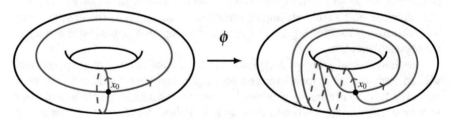

Fig. 2.3 The action on loops of a mapping of the torus

intersections cannot change since the map is one-to-one. This picture illustrates the induced action of the map ϕ on the fundamental group of the torus, $\pi_1(T^2,x_0)$:

$$\phi : T^2 \to T^2 \qquad \text{induces} \qquad \phi_* : \pi_1(T^2,x_0) \to \pi_1(T^2,x_0). \qquad (2.5)$$

As we described in the previous section, the fundamental group contains equivalence classes of loops on a surface, where the equivalence is homotopy of paths (continuous deformations fixing x_0). The closed curves begin and end at x_0, the intersection of the curves in Fig. 2.3. We assume for simplicity that ϕ preserves the basepoint ($\phi(x_0) = x_0$). The fundamental group of the torus is $\mathbb{Z}^2$ [89], so it is actually isomorphic to $H_1(T^2,\mathbb{Z})$, the first homology group with coefficients in $\mathbb{Z}$. Hence, we can represent the action of ϕ_* by a matrix in $SL_2(\mathbb{Z})$. The matrix must have determinant one since ϕ^{-1} must also induce an action ϕ_*^{-1} on $\mathbb{Z}^2$. (A negative determinant is ruled out since ϕ preserves orientation.)

The striking result is that the map

$$\sigma : \text{MCG}(T^2) \to \text{SL}_2(\mathbb{Z}) \qquad (2.6)$$

given by action on first homology is an isomorphism. To put it another way, the mapping class group of the torus is given entirely by the action of diffeomorphisms on $H_1(T^2, \mathbb{Z})$. Indeed, the map σ is well-defined: given a mapping $\phi \in \text{Diff}^+(T^2)$, any other mapping isotopic to ϕ will act on $H_1(T^2, \mathbb{Z})$ with a matrix $\chi_* \circ \phi_*$, where $\chi_* \in \text{Diff}_0(T^2)$. But $\chi_* = \text{id}$ since the isotopy used to take χ back to the identity map will also take loops back to their initial state, leaving any loop unchanged. Hence, all elements of the isotopy class of ϕ act the same way on $H_1(T^2, \mathbb{Z})$.

The map σ is also surjective. Given any element M of $\text{SL}_2(\mathbb{Z})$, we can construct an orientation-preserving linear diffeomorphism of $\mathbb{R}^2$. Taking the torus as the unit square with edges identified, we can see that this linear map induces a diffeomorphism of the torus. (This requires showing equivariance with respect to the deck transformation of the covering space.) This diffeomorphism induces an action on $H_1(T^2, \mathbb{Z})$ with matrix M.

The injectivity of σ is the harder part to prove: to show that two non-isotopic maps must act differently on $H_1(T^2, \mathbb{Z})$. Since this is more technical, we refer the reader to Farb and Margalit [45, p. 53].

2.4 Classification of MCG(T^2)

In Sect. 2.3, we discussed the isomorphism σ that takes an element of MCG(T^2) to a matrix in $\text{SL}_2(\mathbb{Z})$. For definiteness, we write an arbitrary element M of $\text{SL}_2(\mathbb{Z})$ as

$$M = \begin{pmatrix} a & b \\ c & d \end{pmatrix}, \qquad ad - bc = 1. \qquad (2.7)$$

The characteristic polynomial of M is

$$p(x) = x^2 - \tau x + 1, \qquad \tau := a + d, \qquad (2.8)$$

where τ is the trace of M. We will now classify the possible types of mapping classes by examining how powers of M behave. We will make use of the characteristic polynomial $p(x)$.

The eigenvalues of M are $x_\pm = \frac{1}{2}(\tau \pm \sqrt{\tau^2 - 4})$ (with $x_+ x_- = 1$), suggesting that $|\tau| = 2$ plays a special role. In fact, low values of $|\tau|$ are important ($|\tau| = 0, 1, 2$), and we will increase its value gradually in our analysis. We now proceed to investigate the different varieties of matrices M that we can encounter: elliptic, parabolic, and hyperbolic.

2.4.1 Elliptic Case

Recall the Cayley–Hamilton theorem, which says that a matrix is a root of its characteristic polynomial: $p(M) = 0$, or from (2.8)

$$p(M) = M^2 - \tau M + I = 0. \tag{2.9}$$

(Constants are interpreted as multiples of the identity matrix, I.)

Consider first $|\tau| < 2$. Since τ is an integer, this means $\tau = -1, 0, 1$. If $\tau = 0$, then by the Cayley–Hamilton theorem (2.9) $M^2 + I = 0$, so that $M^4 = I$. If $\tau = \pm 1$, then by the same theorem $M^2 \mp M + I = 0$, so $M^2 = \pm M - I$. Multiplying both sides by M, we obtain $M^3 = \pm M^2 - M = \pm(\pm M - I) - M = \mp I$. Hence, $M^6 = I$ for $|\tau| = 1$. The least common multiple of 4 and 6 is 12, so we conclude that $M^{12} = I$ for $|\tau| < 2$.

If $|\tau| = 2$ and $b = c = 0$, we have $M = \pm I$, so that $M^2 = I$.

Putting these together, we conclude that if $|\tau| < 2$ or $M = \pm I$, then $M^{12} = I$. This behavior is called *finite-order*, and the matrix M is *elliptic*: after a while (at most 12 applications), powers of M repeat themselves. Since M acts on homological generators, these must all return to their initial configuration. It follows that ϕ^{12} is isotopic to the identity. Note that this does not imply that ϕ^{12} itself is the identity: in fact ϕ could be a complicated map with chaotic orbits. But topologically, it is indistinguishable from the identity.

2.4.2 Parabolic Case

Building on Sect. 2.4.1, let us consider $\tau = \pm 2$, but assume that $|M| \neq I$, where $|M|$ is the elementwise absolute value. The eigenvalues of M are degenerate and both equal to $\tau/2 = \pm 1$. The Cayley–Hamilton theorem says that $(M \mp I)^2 = 0$, or $M = \pm I + N$, where $N \neq 0$ is *nilpotent* ($N^2 = 0$). A matrix M with this property is called *parabolic*.

It is a simple exercise to show that $e = ((a - \frac{1}{2}\tau), c)$ is an eigenvector (making use of $\tau/2 = 2/\tau$), unless $c = 0$ in which case we use instead $e = (b, (d - \frac{1}{2}\tau))$. The matrix M has only one eigenvector.

Thus, M leaves invariant a loop e and its multiples,[1] possibly reversing their orientation if $\tau = -2$. (We can make e *primitive* by dividing by the greatest common divisor of its two entries. This means that e does not trace the same oriented closed curve more than once.) Every other loop is affected by M. This is in contrast to the elliptic (finite-order) case, where either every loop is untouched by M^p, or they all are. The invariant e corresponds to an equivalence class of *reducing curves*.

Two distinguished parabolic elements of $\mathrm{MCG}(T^2)$ play a special role:

---

[1] We really mean an equivalence class of loops here, but from now on we will often not distinguish the two.

$$T_1 = \begin{pmatrix} 1 & 0 \\ 1 & 1 \end{pmatrix}, \qquad T_2 = \begin{pmatrix} 1 & -1 \\ 0 & 1 \end{pmatrix}. \tag{2.10}$$

These have, respectively, a unique eigenvector $(0, 1)$ and $(1, 0)$. The first matrix, when acting on the loop (p, q), gives

$$\begin{pmatrix} 1 & 0 \\ 1 & 1 \end{pmatrix} \begin{pmatrix} p \\ q \end{pmatrix} = \begin{pmatrix} p \\ p+q \end{pmatrix}. \tag{2.11}$$

It has changed the *twist* in the second component (for $p \neq 0$). This action is illustrated in Fig. 2.4, using the standard basis where $(1, 0)$ is a horizontal loop and $(0, 1)$ is a vertical loop. The linear diffeomorphisms constructed from T_1 and T_2 are called *Dehn twists* along the curves $(0, 1)$ and $(1, 0)$, respectively.

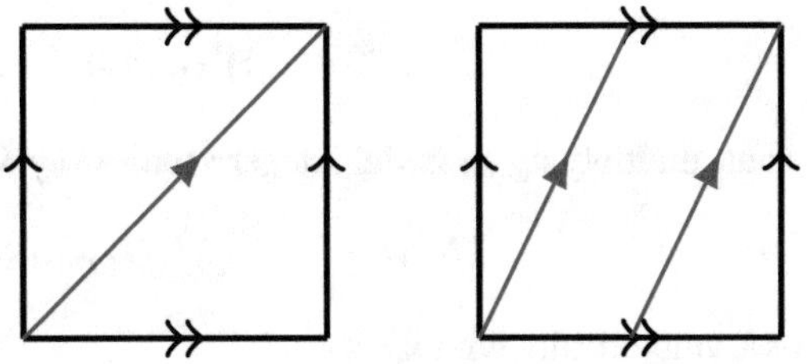

Fig. 2.4 The torus loop $(1, 1)$ (left) and its image $(1, 2)$ (right) under the action of the Dehn twist (2.11)

The two matrices (2.10) *generate* the group $SL_2(\mathbb{Z})$, in the sense that any element can be written as a product of these two matrices and their inverses, possibly repeated.

2.4.3 Hyperbolic Case

Finally, we come to the most interesting case: $|\tau| > 2$. In that case, the two eigenvalues $x_\pm = \frac{1}{2}(\tau \pm \sqrt{\tau^2 - 4})$ are real and distinct. We call $\lambda = \frac{1}{2}(|\tau| + \sqrt{\tau^2 - 4}) > 1$ the *dilatation*[2] of the mapping class corresponding to ϕ. ($\log \lambda$ is called the *topological entropy*.) The two eigenvalues can be written $\pm\{\lambda, \lambda^{-1}\}$, where the sign is the same as τ. It is a standard result that λ is irrational for such *hyperbolic* matrices. For instance, we can easily show that the continued fraction expansion of λ (for $\tau > 2$) is $[\tau - 1; 1, \tau - 2, 1, \tau - 2, \ldots]$, with period 2.

The eigenvectors of M can be written

$$u = \begin{pmatrix} \pm\lambda - d \\ c \end{pmatrix} \quad \text{and} \quad s = \begin{pmatrix} \pm\lambda^{-1} - d \\ c \end{pmatrix}, \tag{2.12}$$

respectively, with eigenvalue $\pm\lambda$ and $\pm\lambda^{-1}$. The slope of these vectors is irrational, which means that no loop remains invariant (or is multiplied by a constant) under

---

[2] This is also called the dilation, stretch factor, expansion constant, or growth.

the action of M. The eigenvector u is called the *unstable direction*, and s the *stable direction*. In the hyperbolic case, the matrix M is always diagonalizable since its eigenvalues are distinct. We can thus write

$$M = \pm(\lambda\, u \otimes U + \lambda^{-1} s \otimes S), \tag{2.13}$$

where $u \otimes U$ denotes the outer product of the two vectors: a rank one matrix with components $(u \otimes U)_{ij} = u_i U_j$. The vectors U and S are the left eigenvectors of M, i.e., the eigenvectors of the transpose of M, with eigenvalue $\pm\lambda$ and $\pm\lambda^{-1}$, respectively. They satisfy $U \cdot u = S \cdot s = 1$, $U \cdot s = S \cdot u = 0$. Observe that the form (2.13) is consistent with $Mu = \pm\lambda u$ and $Ms = \pm\lambda^{-1}s$.

The unstable eigenvector u is exposed by *iteration*, or repeated multiplication by M. Using (2.13) and the orthogonality properties of the left and right eigenvectors, we write the kth power of M as

$$M^k = (\pm 1)^k (\lambda^k u \otimes U + \lambda^{-k} s \otimes S). \tag{2.14}$$

Thus, multiplying an initial integer vector w by M^k gives

$$M^k w = (\pm 1)^k (\lambda^k (U \cdot w) u + \lambda^{-k} (S \cdot w) s) \tag{2.15}$$

or asymptotically with k,

$$M^k w \sim (\pm 1)^k \lambda^k (U \cdot w) u, \qquad k \to \infty, \tag{2.16}$$

since $U \cdot w \neq 0$ because U has irrational slope and w is a vector of integers. We conclude the vector w aligns with the unstable eigenvector u under repeated multiplication by M. Similarly, the vector w aligns with the stable eigenvector s under repeated multiplication by M^{-1}. Interpreting w as some initial loop on the torus, we can see that under iteration by M the loop grows in length and becomes more and more tightly wound around the torus (Fig. 2.5).

The eigenvectors u and s do not themselves correspond to closed loops on the torus. A straight curve through a given point on T^2 parallel to u or s has irrational slope, so that it will wind densely around the torus but never repeat itself. The M-invariant object corresponding to the union of all such curves is called the *unstable foliation*, $\mathscr{F}_u$, when the curves are parallel to the unstable direction (and similarly for the stable foliation, $\mathscr{F}_s$). A single curve in $\mathscr{F}_u$ or $\mathscr{F}_s$ is called a *leaf* of the foliation. Under the action of M, a leaf of $\mathscr{F}_u$ is stretched by λ, while it is contracted by λ^{-1} for $\mathscr{F}_s$.

When we construct a map by taking the linear action of M on $\mathbb{R}^2$ and projecting on the torus, we obtain an *Anosov diffeomorphism*. The most famous example is *Arnold's cat map*, with matrix

$$M = \begin{pmatrix} 2 & 1 \\ 1 & 1 \end{pmatrix}, \tag{2.17}$$

with $\lambda = \frac{1}{2}(3+\sqrt{5}) = 2.61803\ldots$, the square of the Golden Ratio $\varphi = \frac{1}{2}(1+\sqrt{5}) = 1.61803\ldots$. The action of this map on a loop is shown in Fig. 2.5. The number of

Fig. 2.5 The torus loop $(1,1)$ (left) and its images $(3,2)$ (center) and $(8,5)$ (right) under the action of the Anosov (2.17). The slope of the loop rapidly approaches the inverse of the Golden Ratio. In the last frame, the slope is $\frac{5}{8} = 0.625$, whereas $\phi^{-1} = 0.618\ldots$

windings around each direction of the torus increases exponentially. After many iterations, the number of windings is multiplied by λ at each iteration. The curve approaches exponentially a leaf of the unstable foliation, $\mathscr{F}_{u}$.

2.5 Summary

We have introduced several key concepts in this chapter:

- Diffeomorphisms are invertible maps that act smoothly on the torus.
- If we allow continuous deformations of the maps (isotopy), the maps can be classified by examining their action on the first homology group of the torus. This action is linear and is given by 2×2 integer matrices with unit determinant.
- One way to distinguish the maps is to iterate them—to act repeatedly on the same loop. In the elliptic case, the loop does not grow. In the parabolic case, the loop grows linearly under iteration. In the hyperbolic case, the loop grows exponentially with the number of iterations.
- In the hyperbolic case, the growth is characterized by the spectral radius (eigenvalue of the largest magnitude) of a matrix. The value of the spectral radius is called the dilatation of the mapping class.
- We think of the loops as being *stretched* by the map. This is akin to the stretching of taffy discussed in the introduction (Sect. 1.2). One iteration is like a full cycle of the rod motion. This identification will be made more precise in the next chapter.

Iteration is a key concept in dynamics: repeated iteration for a long time is often used to "discover" properties of some system. The long-time limit is also often necessary to prove theorems.

In the next chapter, we shall relate maps of the torus to maps of a disk with 3 punctures. Even though we shall leave behind maps of the torus after this, they are worthy of study in their own right and can serve as a prototype for more complicated behavior [112].

Chapter 3
Stretching with Three Rods

With only two hooks there could be no lapping of the candy, because there was no third pin to re-engage the candy while it was held between the other two pins.
—*Chief Justice W. H. Taft,* Hildreth v. Mastoras

In Chap. 2, we discussed the mapping class group of the torus and showed that it contains three types of elements: periodic, parabolic, and hyperbolic (Anosov). Now we will see how maps of the torus can be associated with maps of the sphere, with some special points. This will open up the analysis of *three-rod stirring devices,* such as the taffy puller shown in Fig. 1.3a–b.

3.1 From the Torus to the Sphere

Consider the linear map $\iota : T^2 \to T^2$, defined by $\iota(x) = -x$, where we regard the torus as the periodic unit square. This map is called the *hyperelliptic involution,* with $\iota^2 = \mathrm{id}$. It has four fixed points:

$$p_0 = \begin{pmatrix} 0 \\ 0 \end{pmatrix}, \qquad p_1 = \begin{pmatrix} \frac{1}{2} \\ 0 \end{pmatrix}, \qquad p_2 = \begin{pmatrix} \frac{1}{2} \\ \frac{1}{2} \end{pmatrix}, \qquad p_3 = \begin{pmatrix} 0 \\ \frac{1}{2} \end{pmatrix}, \qquad (3.1)$$

since $-\frac{1}{2}$ is identified with $\frac{1}{2}$ because of periodicity (i.e., mod 1). These are the *Weierstrass points* of ι, whose name comes from the Weierstrass $\wp$-function, which maps the periodic plane to the sphere. The map ι rotates the unit square by 180° about its middle point $(\frac{1}{2}, \frac{1}{2})$. The map ι is minus the identity matrix and therefore is in the *center* of $\mathrm{MCG}(T^2)$ since it commutes with all its elements.

Now let us construct the quotient space

$$S = T^2/\iota, \qquad (3.2)$$

that is, the torus where points related by the involution are identified with each other. The map $\pi : T^2 \to T^2/\iota$ is the canonical projection. Figure 3.1a shows how the points on the torus are related to each other through ι and the torus edge identifications in Fig. 2.1b. In Fig. 3.1b, we discard a redundant half of the surface. The

J.-L. Thiffeault, *Braids and Dynamics*, Frontiers in Applied Dynamical Systems: Reviews and Tutorials 9, https://doi.org/10.1007/978-3-031-04790-9_3

edge identifications in the figure come from a mix of periodicity of the original torus and the rotation by $180°$ used in constructing the quotient surface.

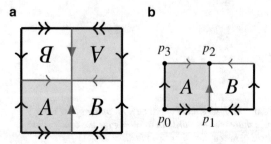

Fig. 3.1 (**a**) Identifications of regions on T^2 under the map ι. The two red lines are invariant. (**b**) The surface $S = T^2/\iota$, with the four Weierstrass points shown

The quotient surface S in Fig. 3.1b is a sphere, which we can show as follows. We zip along the identified edges between p_0 and p_1 and between p_2 and p_3, to obtain the surface in Fig. 3.2a. There is only one identification left, between the left and right boundaries of that surface. If we glue these together, we obtain a surface shaped like an American football, with p_0 and p_3 at the tips. This surface has the same topology as a sphere, with four "distinguished" points corresponding to the Weierstrass points. This final surface is shown in Fig. 3.2b.

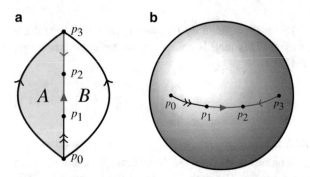

Fig. 3.2 (**a**) Zipped surface. (**b**) $S = T^2/\iota$ is a sphere with four punctures, denoted $S_{0,4}$

The four Weierstrass points can be removed to create punctures on the sphere; we denote the surface in Fig. 3.2b as $S_{0,4}$: a surface of genus 0 with 4 punctures. The interpretation of the four points as punctures comes from the fact that homotopy classes of closed curves on T^2 project down under π to homotopy classes of closed curves on $S_{0,4}$.

In Fig. 3.3, we show two closed curves α and β on T^2 that intersect once. To project the curves to the punctured sphere $S_{0,4}$, we first pair each curve with its ι-image, to obtain $\alpha \cup \iota\alpha$ and $\beta \cup \iota\beta$ (Fig. 3.3, middle). These paired curves are ι-invariant and so may be projected down to S without ambiguity. In Fig. 3.3 (right), we then take some arbitrary half of the torus to represent the quotient surface S, as we did in Fig. 3.1b. The projection of the curves $\alpha \cup \iota\alpha$ and $\beta \cup \iota\beta$ on $S_{0,4}$ yields two closed curves that intersect twice (Fig. 3.4).

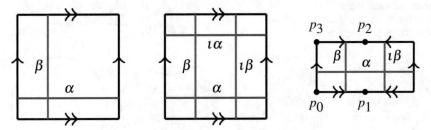

Fig. 3.3 Two intersecting closed curves on the torus (left) and their image under ι (middle). The curves and their ι-images project down to the quotient surface $S = T^2/\iota$ (right)

In fact, there is a bijection between homotopy classes of closed curves on T^2 and those on $S_{0,4}$ [45, p. 55]. As we just saw, the preimage of a closed curve on $S_{0,4}$ is a "double closed curve" on T^2 since given a curve α on T^2, both α and $\iota\alpha$ map to the same curve on $S_{0,4}$. Note that this construction does not work for oriented curves since the preimage of an oriented curve on $S_{0,4}$ gives two disjoint identical curves with opposite orientation on T^2 (related by ι) and therefore does not give a well-defined homotopy class.

3.2 The Mapping Class Groups of $S_{0,4}$ and D_3

Mapping classes on T^2 represented by ϕ and $\iota\phi$ project down to a well-defined mapping class on $S_{0,4}$. The elements of this group are in $\mathrm{SL}_2(\mathbb{Z})/\iota =: \mathrm{PSL}_2(\mathbb{Z})$, the *projective* version of the linear group. This means that the mapping class group of $S_{0,4}$ has "fewer" elements than that of T^2, but it gains a few more because it is possible to interchange the four punctures on the sphere pairwise through rotation, which lifts to the identity map on the (unpunctured) torus. These rotations by $180°$ are the two hyperelliptic involutions of $S_{0,4}$. The full mapping class group is thus [45, p. 56][1]

$$\mathrm{MCG}(S_{0,4}) \approx \mathrm{PSL}_2(\mathbb{Z}) \ltimes (\mathbb{Z}_2 \times \mathbb{Z}_2). \tag{3.3}$$

The operator $\ltimes$ is a *semidirect product* of the group $\mathrm{PSL}_2(\mathbb{Z})$ acting on the group $\mathbb{Z}_2 \times \mathbb{Z}_2$, the latter being the group generated by the two hyperelliptic involutions of $S_{0,4}$.

We shall not dwell on this semidirect product structure because for applications we are actually interested in a slightly different surface: D_3, the *disk* with 3 punctures. This is topologically the same as $S_{0,4}$ if we remove a small disk around one of the punctures and then stretch this disk to make the outer boundary of D_3 (Fig. 3.5).

---

[1] The structure is semidirect since an element of $\mathrm{PSL}_2(\mathbb{Z})$ acts on $\mathbb{Z}_2 \times \mathbb{Z}_2$ by permuting the punctures.

Fig. 3.4 The image of the
two closed curves in Fig. 3.3
on the punctured sphere $S_{0,4}$
(sphere not shown). The
curves intersect twice.
Here, $\pi : T^2 \to S_{0,4}$ is the
canonical projection map

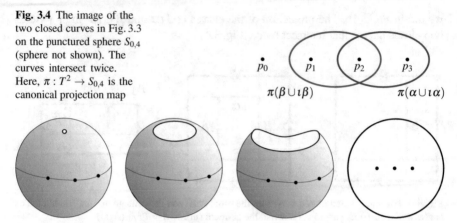

Fig. 3.5 After removing a small disk around one of the punctures (left), $S_{0,4}$ is turned into the disk D_3 by stretching the boundary (right)

A consequence of replacing a puncture by a boundary is that it is no longer equivalent to the other punctures, so diffeomorphisms cannot interchange the boundary with any puncture. This rules out the hyperelliptic involutions of $S_{0,4}$ as allowable diffeomorphisms, so we lose the factor $\mathbb{Z}_2 \times \mathbb{Z}_2$ in Eq. (3.3).

To fully characterize $\mathrm{MCG}(D_3)$, we must specify how diffeomorphisms of D_3 to itself treat the boundary, ∂D. There are two natural options: fix ∂D pointwise or setwise. If we fix it pointwise, a diffeomorphism can "twist" around the boundary by rotating it by multiples of 2π. Isotopies also fix the boundary pointwise, so these cannot be untwisted. The mapping class group will then contain these rotations and will have a direct product group structure:

$$\mathrm{MCG}(D_3; \partial D \text{ fixed pointwise}) \approx \mathrm{PSL}_2(\mathbb{Z}) \times \mathbb{Z}. \qquad (3.4)$$

The factor $\times\mathbb{Z}$ counts the number of twists of the boundary. We will elaborate more on this direct product structure at the end of Sect. 4.2.

If we fix ∂D setwise, then isotopies can freely rotate ∂D and remove any twist imposed by the diffeomorphism. This is the same as if the boundary was a puncture, so we obtain

$$\mathrm{MCG}(D_3; \partial D \text{ fixed setwise}) \approx \mathrm{PSL}_2(\mathbb{Z}). \qquad (3.5)$$

We shall usually assume that boundaries are fixed setwise, not pointwise, unless otherwise noted.

A comment is in order when dealing with fluid-mechanical applications. When the fluid is viscous (i.e., obeys the Navier–Stokes equation), it is natural to fix the boundary pointwise since the fluid is not allowed to "slip" along the boundary. When the fluid is inviscid (i.e., obeys Euler's equation), it can slip and the setwise defi-

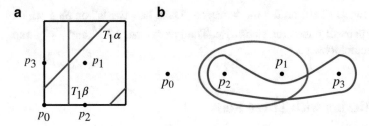

Fig. 3.6 (a) The action of T_1 on the two curves from Fig. 3.3. (b) The induced action of T_1, denoted $\overline{T}_1$, on the two curves from Fig. 3.4

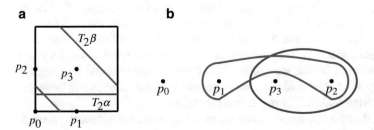

Fig. 3.7 (a) The action of T_2 on the two curves from Fig. 3.3. (b) The induced action of T_2, denoted $\overline{T}_2$, on the two curves from Fig. 3.4

nition is more appropriate. However, both definitions give the same answer when dealing with growth of curves.

3.3 Dehn Twists

Recall the two Dehn twists on the torus, Eq. (2.10): how do they act on the punctures of $S_{0,4}$? We use the same symbol for the mapping class and the linear diffeomorphism constructed from T_i and regard the T_i as acting on $S_{0,4}$ via the quotient by the hyperelliptic involution, as described in Sect. 3.1. We write $\overline{T}_i = \pi \circ T_i \circ \pi^{-1}$ for the induced action on $S_{0,4}$. These maps are well-defined since the T_is commute with the hyperelliptic involution.

Clearly, $T_1 p_0 = T_2 p_0 = p_0$, and it is easy to check

$$
\begin{aligned}
T_1(p_1) &= p_2, & T_2(p_1) &= p_1, \\
T_1(p_2) &= p_1, & T_2(p_2) &= p_3, \\
T_1(p_3) &= p_3, & T_2(p_3) &= p_2.
\end{aligned}
$$

The action of T_1 on the two closed curves from Figs. 3.3 and 3.4 is shown in Fig. 3.6a–b. The effect of T_1 on $S_{0,4}$ is to interchange p_1 and p_2 clockwise: the red curve is invariant since it contains p_1 and p_2; the blue curve is pulled and stretched below p_1 by the motion of p_2. Figure 3.7a–b show the action of T_2 on the same curves: this time p_2 and p_3 are interchanged clockwise on $S_{0,4}$. We conclude that the two Dehn twists (2.10) on the torus map to "Dehn half-twists" on $S_{0,4}$, which swap adjacent pairs of punctures clockwise. The inverse maps $\overline{T}_1^{-1}$ and $\overline{T}_2^{-1}$ swap punctures counterclockwise.

3.4 Fluid Stirring with Three Rods

A picture such as the rightmost frame in Fig. 3.5 can be interpreted as a circular container of fluid, with three movable stirring rods immersed [27, 121]. We neglect fluid motion perpendicular to the page, which is a good approximation when the fluid layer is thin, or stratified because of thermal effects. Now move the rods in some periodic fashion, so that they return to their initial positions as a set (they might have been permuted). The rods displace the fluid, which induces a diffeomorphism of D_3 to itself, which we call ϕ_1. This diffeomorphism could be computed numerically by solving a set of equations appropriate to the fluid: Stokes (slow viscous flow), Navier–Stokes, Oldroyd B (polymers), etc.

Now move the rods again in the same manner. This induces a diffeomorphism ϕ_2, which is not in general the same as ϕ_1: for instance, the system might be getting more and more turbulent at each period, such as when we stir cream into coffee.[2] However, they must be isotopic, $\phi_1 \simeq \phi_2$, since the rods were moved the same way both times and the rods' periodic motion completely determines the mapping class (an element of $\mathrm{MCG}(D_3) \approx \mathrm{PSL}_2(\mathbb{Z})$, see Sect. 3.2). This means that we can use any of these maps, say ϕ_1, to represent the element of the mapping class $[\phi]$ induced by the rod motion.

Let us select a way of moving the rods by choosing an element of $\mathrm{MCG}(D_3)$. This corresponds to the isotopy class of a linear diffeomorphism $\phi : T^2 \to T^2$, which projects down to a diffeomorphism $\pi \circ \phi \circ \pi^{-1} = \bar{\phi} : D_3 \to D_3$ using the construction in Sect. 3.1. This mapping is well-defined since linear diffeomorphisms on T^2 commute with the hyperelliptic involution ι. A good way of characterizing ϕ is to write it as a sequence of Dehn twists T_1 and T_2. Each of these Dehn twists maps to an interchange of punctures, as described in Sect. 3.3.

Let us illustrate this with a simple example that is classical but was first discussed in the context of fluid dynamics by Boyland et al. [27] (see Fig. 1.2, as well as Fig. 1.4 for the rod motion). We start with an interchange of the first and second rods clockwise ($\overline{T}_1$) and then interchange the second and third rods counterclockwise ($\overline{T}_2^{-1}$). The net result is the map $\bar{\phi} = \overline{T}_2^{-1} \overline{T}_1$. This lifts to a linear map on the torus $\phi = T_2^{-1} T_1$. The matrix representing this map is the Arnold cat map matrix (2.17), so we know the isotopy class of ϕ is Anosov (hyperbolic). The induced isotopy class of $\bar{\phi}$ on D_3 is called *pseudo-Anosov*; for reasons, we will explain in Chap. 5.

---

[2] Studies of topological mixing are usually in the limit of zero Reynolds number, so the fluid motion is exactly periodic. Finn et al. [52] and Smith and Warrier [109] have studied the eddies shed from rods at moderate Reynolds number from a topological perspective.

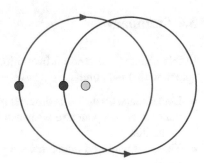

Fig. 3.8 The orbits of the rods
in the taffy puller in Fig. 1.3b

What can we deduce about the action of $\bar{\phi}$? First, since ϕ is Anosov, any essential simple closed curve will asymptotically grow exponentially under repeated action of ϕ. But since essential simple closed curves on T^2 map to such curves on D_3 (see Figs. 3.3 and 3.4), we will observe exponential growth on D_3 as well. Thus, the fluid motion on D_3 carries at least the same "complexity" as that on T^2. This can be made rigorous in terms of semiconjugacy of maps [27, 67]. Tumasz and Thiffeault [128, 129] have studied the gap in complexity between the fluid motion and the rod motion, as measured by the topological entropy. The idea that a clever choice of rod motion can guarantee a certain level of complexity in the fluid has been used for designing fluid mixing devices [16, 16, 26, 50, 60, 74, 111].

3.5 Taffy Pulling with Three Rods

The motion of the rods in the taffy puller of Fig. 1.3b is shown in Fig. 3.8. The two moving rods never collide. During their motion, they only exchange position with the fixed rod. After a bit of staring at Fig. 4.2a, we see that the rods induce the map $\overline{T}_2^{-1}\overline{T}_1^2\overline{T}_2^{-1}$, where the matrices are defined in (1.1). (This map is associated with the braid $\sigma_2^{-1}\sigma_1^2\sigma_2^{-1}$, as we shall see in the next chapter; see Fig. 4.2a–b.) Here there is no "physical" ambient space: we simply imagine that the rods are puncturing a hypothetical disk, which we use to measure the taffy's stretching motion. The linear diffeomorphism on T^2 corresponding to the rod motion is

$$\overline{T}_2^{-1}\overline{T}_1^2\overline{T}_2^{-1} = \begin{pmatrix} 1 & 1 \\ 0 & 1 \end{pmatrix} \begin{pmatrix} 1 & 0 \\ 2 & 1 \end{pmatrix} \begin{pmatrix} 1 & 1 \\ 0 & 1 \end{pmatrix} = \begin{pmatrix} 3 & 4 \\ 2 & 3 \end{pmatrix} \tag{3.6}$$

with dilatation $\lambda = 3 + 2\sqrt{2}$. This gives us the factor by which the length of the taffy is multiplied at each full period. In that sense, it is a measure of the effectiveness of the taffy puller. Thiffeault [118] compared various taffy puller designs by computing their associated dilatation. The number $3 + 2\sqrt{2} = (1 + \sqrt{2})^2$ recurs several times in these designs—it is the *silver ratio* [50], a close relative of the Golden ratio.

3.6 Summary

In this chapter, we used an involution of the torus to create a quotient space, the sphere with 4 punctures.

- Linear maps on the torus descend nicely to diffeomorphisms of the sphere with 4 punctures. One puncture is always fixed, so we interpret it as the outer boundary of the disk D_3.
- These remaining 3 punctures are possibly permuted by the induced map on the sphere.
- The parabolic matrices acting on the torus are Dehn half-twists on the disk. These elementary operations swap adjacent punctures, either clockwise or counterclockwise. They can be used as building blocks for any mapping class of the punctured disk D_3.
- The mapping class group of D_3 is $\mathrm{PSL}_2(\mathbb{Z})$ when the boundary is fixed setwise. This is the projective version of $\mathrm{SL}_2(\mathbb{Z})$, where matrices M and $-M$ are identified.
- The mapping class group of D_3 is $\mathrm{PSL}_2(\mathbb{Z}) \times \mathbb{Z}$ when the boundary is fixed pointwise. The extra factor of $\mathbb{Z}$ counts how many times the outer boundary was twisted by the diffeomorphism.
- We can identify the 3 moving punctures with the rods in a stirring device or taffy puller. In that case, the dilatation (growth) is proportional to the efficiency of the mixing device or taffy puller [26, 50, 118].

Everything in this chapter applied to 3 and only 3 punctures. In that case, the dynamics on the torus perfectly capture the dynamics on D_3. If we want to allow for more punctures in our dynamics, we first need a compact way of describing an arbitrary mapping class of the disk D_n. We shall do so in the next chapter by introducing the braid group B_n.

Chapter 4
Braids

> *Although it has been proved that every braid can be deformed into a similar normal form the writer is convinced that any attempt to carry this out on a living person would only lead to violent protests and discrimination against mathematics. He would therefore discourage such an experiment.*
>
> — *Emil Artin,* Theory of Braids [8]

At the end of Chap. 3, we studied the motion of rods defining a 3-rod mixing device or taffy puller. We then associated these motions with the Dehn half-twists $\overline{T}_1$ and $\overline{T}_2$ that describe the clockwise interchange of punctures. If we want to study motions involving more rods, it is natural to introduce the *braid group* B_n for some fixed number of punctures n. This will allow us to encode much more complex motions.

4.1 Braids as Particle Dances

Artin [7] first introduced braids as collections of strings connecting points. He gave intuitive justification for several theorems and in [8] provided rigorous proofs. There are several ways to define braids [18, 19, 104], but here we use the "braids as particle dances" viewpoint since it is intimately connected to dynamics. Consider n particles (or punctures) located at points in the Euclidean plane $\mathbb{E}^2$, which we regard as points in the complex plane $\mathbb{C}$. Now assume the points can move,

$$z(t) = (z_1(t), \ldots, z_n(t)), \qquad t \in \mathbb{I} = [0, 1], \tag{4.1}$$

such that

$$z_j(t) \neq z_k(t), \qquad j \neq k. \tag{4.2}$$

These last conditions mean that the particles never collide during their motion. Furthermore, assume that the points collectively return to their initial position:

$$\{z_1(0), \ldots, z_n(0)\} = \{z_1(1), \ldots, z_n(1)\} \qquad \text{(setwise).} \tag{4.3}$$

The specification "setwise" is important since it means the particles can be permuted among themselves. Such a vector of trajectories $z(t)$ defines a *braid*. If we

© The Author(s), under exclusive license to Springer Nature Switzerland AG 2022
J.-L. Thiffeault, *Braids and Dynamics*, Frontiers in Applied Dynamical Systems:
Reviews and Tutorials 9, https://doi.org/10.1007/978-3-031-04790-9_4

embed $\mathbb{E}^2 \times \mathbb{I}$ in $\mathbb{E}^3$ (what physicists call a space–time diagram), we obtain *geometric* or *physical* braids (Fig. 4.1a).

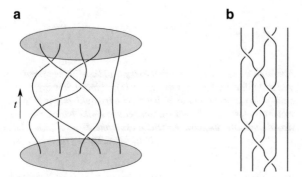

a   **b**

Fig. 4.1 (**a**) A geometric braid. (**b**) The standard braid diagram corresponding to the braid on the left

Intuitively, we claim that two braids are equal if they can be deformed into each other without any strings crossing. This leads to the notion of equivalence, or isotopy, of braids. Two braids are equivalent if there exists a homotopy from the first braid to the second, fixing the endpoints, such that at any time during the deformation the homotopy gives a valid braid [18]. Since we will essentially be interested only in equivalence classes of braids under isotopy, and not in specific physical braids, from now on we drop "equivalence class" and assume that equality of braids means equivalence under isotopy.

Now consider two braids γ and γ', with coordinate vectors $z(t)$ and $z'(t)$, sharing the same endpoints. We define the composition $\gamma'' = \gamma\gamma'$ as the braid with coordinate vector

$$z''(t) = \begin{cases} z(2t), & 0 \leq t \leq \frac{1}{2}; \\ z'(2t-1), & \frac{1}{2} \leq t \leq 1. \end{cases} \tag{4.4}$$

With this composition law, braids form a group B_n, the *Artin braid group on n strings*. It is easy to see that this composition law is associative. The identity braid has coordinate vector $z(t) = (z_1(0), \ldots, z_n(0))$. The inverse of γ is the braid γ^{-1} with coordinates $z(1-t)$. All the group axioms are thus satisfied.

4.2 Algebraic Braids

It is common to redraw a braid as a *standard braid diagram*, where the braid is "flattened" and where crossings occur one at a time in the intervals of the same length (Fig. 4.1b). There are various conventions for how to plot these, but here we read the "time" parameter of the braid as flowing from bottom to top, in the same

a   **b**

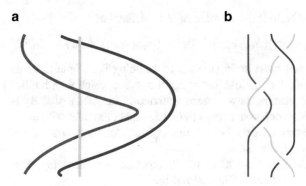

Fig. 4.2 (**a**) The geometric braid associated with the rod motion in Fig. 3.8. (**b**) The standard braid diagram corresponding to the braid $\sigma_2^{-1}\sigma_1^2\sigma_2^{-1}$ on the left

a   **b**

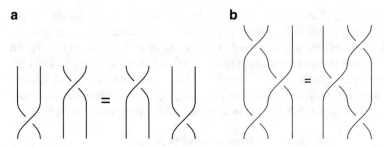

Fig. 4.3 (**a**) Relation $\sigma_j\sigma_k = \sigma_k\sigma_j$. (**b**) Relation $\sigma_j\sigma_k\sigma_j = \sigma_k\sigma_j\sigma_k$

manner as a space–time diagram. Figures 4.2a and b show the physical braid and corresponding standard diagram for the taffy puller in Fig. 1.3.

Figure 4.1b suggests that we rewrite a braid in terms of *standard generators*, denoted σ_i for $i = 1,\ldots,n-1$. The generator σ_i is the clockwise interchange of the ith string with the $(i+1)$th string, and σ_i^{-1} is the corresponding counterclockwise exchange. Any braid can be written as a product of these standard generators and their inverses. They satisfy the relations

$$\sigma_j\sigma_k = \sigma_k\sigma_j, \quad |j-k| > 1; \qquad \sigma_j\sigma_k\sigma_j = \sigma_k\sigma_j\sigma_k, \quad |j-k| = 1. \qquad (4.5)$$

These relations are depicted in Figs. 4.3a–b. The first type of relation says that generators that do not share a string commute. The second is less obvious and involves triplets of adjacent strings; staring at Fig. 4.3b for a while should convince the reader that the relations hold. The latter are the key defining relations for the braid group and are often referred as the *braid relations* when they hold, even for groups other than the braid group. Artin [8] proved the somewhat surprising fact that (4.5) are the only nontrivial relations for the braid group B_n. A braid given by a product of generators obeying (4.5) is called an *algebraic braid* since it is not necessarily associated with a physical braid—it may be used as an abstract object.

The (positive) *half-twist* braid is the element $\Delta_n \in B_n$ defined as

$$\Delta_n = (\sigma_{n-1}\sigma_{n-2}\cdots\sigma_1)(\sigma_{n-1}\sigma_{n-2}\cdots\sigma_2)\cdots(\sigma_{n-1}\sigma_{n-2})(\sigma_{n-1}). \qquad (4.6)$$

As its name implies, the half-twist braid consists of grabbing the whole braid as a ribbon and giving it a 180° twist. Note that $\sigma_j\Delta_n = \Delta_n\sigma_{n-j}$, which then implies that $\sigma_j\Delta_n^2 = \Delta_n^2\sigma_j$, i.e., Δ_n^2 commutes with each generator. This means that Δ_n^2 is in the *center* of B_n—the subgroup consisting of elements that commute with every element of B_n. In fact, the center of the braid group is generated by powers of Δ_n^2, which is called the (positive) *full-twist*.

The braid group is most important to us for its connection to mapping class groups of punctured disks and spheres. The claim is that

$$\mathrm{MCG}(D_n; \partial D \text{ fixed pointwise}) \approx B_n; \qquad (4.7)$$

that is, the mapping class group of the disk with n punctures is isomorphic to B_n. Here the boundary of the disk is fixed pointwise by the diffeomorphisms and isotopies. The isomorphism (4.7) is proved explicitly by Birman and Brendle [19]. An intuitive viewpoint is that a braid drags along a "rubber sheet" as it travels from the bottom to the top endpoints. The deformation of the rubber sheet, which is anchored to the boundary, is the diffeomorphism whose mapping class is labeled by the braid.

If the boundary of the disk is fixed as a set rather than pointwise, we have

$$\mathrm{MCG}(D_n; \partial D \text{ fixed setwise}) \approx B_n/\langle\Delta_n^2\rangle. \qquad (4.8)$$

In this case, the diffeomorphisms can "rotate" the boundary, so the isotopy can unwind any net full-twist, hence the quotient by the center that appears in (4.8). Comparing this to (3.5), we see that $B_3/\langle\Delta_3^2\rangle \approx \mathrm{PSL}_2(\mathbb{Z})$.

We finish this section with a technical remark regarding the direct product structure of the mapping class group for the disk with 3 punctures, with boundary fixed pointwise. Comparing (4.7) to (3.4), we can see that $B_3 \approx \mathrm{PSL}_2(\mathbb{Z}) \times \mathbb{Z}$, where recall that the factor $\times\mathbb{Z}$ counts the number of full-twists. The structure is that of a direct product since there is only one boundary, which cannot be permuted by an element of $\mathrm{PSL}_2(\mathbb{Z})$ fixing the boundary. More precisely, there is an exact sequence

$$1 \longrightarrow \mathbb{Z} \longrightarrow B_3 \longrightarrow \mathrm{PSL}_2(\mathbb{Z}) \longrightarrow 1, \qquad (4.9)$$

where $\mathbb{Z}$ is the center of the braid group B_3. The homomorphism from $\mathrm{PSL}_2(\mathbb{Z})$ to B_3 is defined by writing an element of $\mathrm{PSL}_2(\mathbb{Z})$ in terms of the presentation $\langle a, b | aba = bab, (ab)^3 = 1\rangle$ and then interpreting a and b in terms of the presentation of $B_3 = \langle a, b | aba = bab\rangle$. This is well-defined since $(ab)^3$ is in the center of B_3. We conclude that the structure is semidirect. But because $\mathrm{PSL}_2(\mathbb{Z})$ acts on $\mathbb{Z}$ by conjugation, and $\mathbb{Z}$ is the center, the action is trivial, and we get a direct product. In other words, B_3 is the universal central extension of $\mathrm{SL}_2(\mathbb{Z})$ [45, p. 87–88].

4.3 Artin's Representation

Consider the generating set of loops for the fundamental group $\pi_1(D_n, x_0)$ in Fig. 4.4. Unlike for the torus case (Chap. 2), the set $\langle e_1, \cdots, e_n \rangle$ generates a *free*

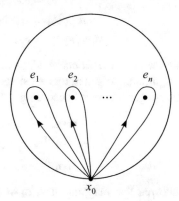

Fig. 4.4 A generating set of loops for $\pi_1(D_n, x_0)$

group, that is, $\pi_1(D_n, x_0) \approx F_n$. The free group with n generators F_n is a finitely generated group where the generators obey no nontrivial relations. (The braid group B_n is not free since its generators satisfy the relations (4.5).) Free groups have the very useful property that two reduced words (i.e., products of generators with no adjacent canceling generators) are equal if and only if they are equal generator-by-generator (also called lexicographical equality).

Recall that the generator $\sigma_i \in B_n$ exchanges the position of the ith and $(i+1)$th punctures. Since elements of B_n correspond to elements of $\mathrm{MCG}(D_n)$ (see (4.7)–(4.8)), there is a canonical diffeomorphism corresponding to σ_i whose induced action on loops is pictured in Fig. 4.5. The loops e_j for $j \neq i, i+1$ are untouched.

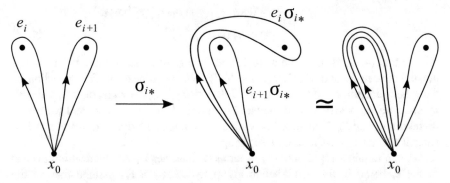

Fig. 4.5 The diffeomorphism corresponding to σ_i acts on the generating loops

We write $\sigma_{i*} : \pi_1(D_n, x_0) \to \pi_1(D_n, x_0)$ for the induced action of σ_i. In mathematical terms, $\sigma_{i*} \in \text{Aut}(F_n)$, the group of *automorphisms* of the free group. The rightmost picture in Fig. 4.5 shows how to express the result of σ_{i*} in terms of the initial loops:

$$e_i \sigma_{i*} = e_i e_{i+1} e_i^{-1}, \tag{4.10a}$$

$$e_{i+1} \sigma_{i*} = e_i, \tag{4.10b}$$

$$e_j \sigma_{i*} = e_j, \qquad j \neq i, i+1. \tag{4.10c}$$

Note that we use a *right action* since in our convention braid words are read from left to right, and so generators must act to their left to maintain the correct order. The inverse generators act according to

$$e_i \sigma_{i*}^{-1} = e_{i+1}, \tag{4.11a}$$

$$e_{i+1} \sigma_{i*}^{-1} = e_{i+1}^{-1} e_i e_{i+1}, \tag{4.11b}$$

$$e_j \sigma_{i*}^{-1} = e_j, \qquad j \neq i, i+1. \tag{4.11c}$$

Artin showed that mapping each σ_i to σ_{i*} defines a faithful representation of the braid group B_n. For instance,

$$
\begin{aligned}
e_i \sigma_{i*} \sigma_{i+1*} \sigma_{i*} &= \left(e_i e_{i+1} e_i^{-1}\right) \sigma_{i+1*} \sigma_{i*} \\
&= \left(e_i \sigma_{i+1*}\right)\left(e_{i+1} \sigma_{i+1*}\right)\left(e_i^{-1} \sigma_{i+1*}\right) \sigma_{i*} \\
&= \left(e_i e_{i+1} e_{i+2} e_{i+1}^{-1} e_i^{-1}\right) \sigma_{i*} \\
&= e_i e_{i+1} e_i^{-1} \, e_i e_{i+2} e_i^{-1} \, e_i e_{i+1}^{-1} e_i^{-1} \\
&= e_i e_{i+1} e_{i+2} e_{i+1}^{-1} e_i^{-1}, \tag{4.12}
\end{aligned}
$$

and

$$
\begin{aligned}
e_i \sigma_{i+1*} \sigma_{i*} \sigma_{i+1*} &= e_i \sigma_{i*} \sigma_{i+1*} \\
&= e_i e_{i+1} e_i^{-1} \sigma_{i+1*} \\
&= \left(e_i \sigma_{i+1*}\right)\left(e_{i+1} \sigma_{i+1*}\right)\left(e_i^{-1} \sigma_{i+1*}\right) \\
&= e_i e_{i+1} e_{i+2} e_{i+1}^{-1} e_i^{-1}. \tag{4.13}
\end{aligned}
$$

We see that (4.12) is the same as (4.13), as they must be since there is an isomorphism between B_n and $\text{MCG}(D_n)$ (see (4.7)). We can repeat this for the action on the other generating loops, and we find that $\sigma_{i*} \sigma_{i+1*} \sigma_{i*} = \sigma_{i+1*} \sigma_{i*} \sigma_{i+1*}$. In this manner, we can verify that all the braid relations (4.5) are satisfied, and we have a representation of B_n. Note that this is not a linear representation in terms of matrices since it acts on a (non-Abelian) free group.

We can obtain a representation in terms of matrices by Abelianization: instead of the e_i being in the non-Abelian group $\pi_1(D_n, x_0) \approx F_n$, assume that all the generators commute.[1] This is a standard procedure that maps the first homotopy

---

[1] This is the same as taking the quotient of F_n with its *commutator subgroup* $[F_n, F_n]$, which consists of all elements that can be written in the form $\alpha \beta \alpha^{-1} \beta^{-1}$, for some $\alpha, \beta \in F_n$.

group $\pi_1(D_n, x_0)$ to the first homology group with integer coefficients, $H_1(D_n, \mathbb{Z}) \approx \mathbb{Z}^n$. Thus, an element $e_1 e_2^2 e_4^{-1} \in \pi_1(D_4, x_0)$ will be written additively as $E_1 + 2E_2 - E_4 \in H_1(D_4, \mathbb{Z})$. The E_i forms a basis for $H_1(D_n, \mathbb{Z})$.

Now apply the Abelianization procedure to the action (4.10). The only nontrivial actions are $E_i \sigma_{i*} = E_{i+1}$ and $E_{i+1} \sigma_{i*} = E_i$. This is simply the permutation of E_i and E_{i+1}: we have recovered a representation for the symmetric group on n symbols in terms of elementary permutation matrices. This is not very useful: it captures nothing of the special character of the braid group. We shall see how to improve this below.

Let us summarize: the representation of B_n in terms of automorphisms $\mathrm{Aut}(F_n)$ is faithful, meaning that it is the identity if and only if the corresponding braid is the identity. However, such a representation is hard to manipulate, as is obvious from (4.12)–(4.13). In particular, iteration (finding the representation of a power of a braid) is difficult. Abelianizing gives us matrices acting on $H_1(D_n, \mathbb{Z})$, which are very easy to multiply together. However, the resulting representation can only characterize the permutation induced by a braid, so the representation is extremely unfaithful and is a poor characterization of the braid group B_n. Is there a way to have our cake and eat it too: to have a representation in terms of matrices that is also faithful? We shall see in Sect. 4.5 that this is (almost) possible. But first we must digress a bit.

4.4 Free Homotopy Representation

We introduce the new loops

$$u_i = e_i e_{i+1}^{-1}, \qquad 1 \le i \le n-1. \tag{4.14}$$

Using (4.10), we compute the action of σ_{i*} on these new loops. We have $u_j \sigma_{i*} = u_j$ for $j < i-1$ or $j > i+1$. The nontrivial actions are

$$\begin{aligned}
u_{i-1} \sigma_{i*} &= (e_{i-1}\sigma_{i*})(e_i^{-1}\sigma_{i*}) \\
&= (e_{i-1})(e_i e_{i+1}^{-1} e_i^{-1}) \\
&= u_{i-1} e_i u_i e_i^{-1},
\end{aligned} \tag{4.15}$$

$$\begin{aligned}
u_i \sigma_{i*} &= (e_i \sigma_{i*})(e_{i+1}^{-1}\sigma_{i*}) \\
&= (e_i e_{i+1} e_i^{-1})(e_i^{-1}) \\
&= e_i u_i^{-1} e_i^{-1},
\end{aligned} \tag{4.16}$$

and

$$u_{i+1}\sigma_{i*} = (e_{i+1}\sigma_{i*})(e_{i+2}^{-1}\sigma_{i*})$$
$$= (e_i)(e_{i+2}^{-1}) \tag{4.17}$$
$$= u_i u_{i+1}.$$

Collecting these results, we have the action

$$u_j\sigma_{i*} = \begin{cases} u_{i-1}e_i u_i e_i^{-1}, & j = i-1; \\ e_i u_i^{-1} e_i^{-1}, & j = i; \\ u_i u_{i+1}, & j = i+1; \\ u_j, & \text{otherwise.} \end{cases} \tag{4.18}$$

Its inverse follows:

$$u_j\sigma_{i*}^{-1} = \begin{cases} u_{i-1}u_i, & j = i-1; \\ e_{i+1}^{-1}u_i^{-1}e_{i+1}, & j = i; \\ e_{i+1}^{-1}u_i e_{i+1}u_{i+1}, & j = i+1; \\ u_j, & \text{otherwise.} \end{cases} \tag{4.19}$$

The presence of e_is and e_{i+1}s in (4.18)–(4.19) means that the action does not close in terms of the u_i. Abelianizing leads to the cancellation of the e_i, and we then obtain matrices that are the analog of the permutation matrices we obtained from Abelianization in the previous section. These matrices are a special case (with $t = 1$) of the reduced Burau representation that we will derive in Sect. 4.5, so we do not describe them here.

The e_i appears only as conjugation $e_i u_i^{\pm 1} e_i^{-1}$ in (4.18), and similarly for (4.18) with e_{i+1}. This suggests passing to *free homotopy* of loops. Free homotopy is essentially homotopy without a basepoint: we are free to deform loops as we please, as long as we do not cross the punctures in D_n. The loop u_i is depicted in Fig. 4.6, be-

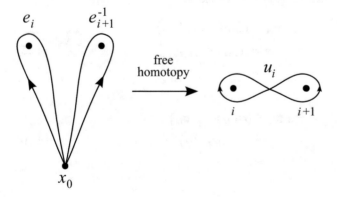

Fig. 4.6 The loop $u_i = e_i e_{i+1}^{-1}$ after passing to free homotopy by relaxing the basepoint x_0

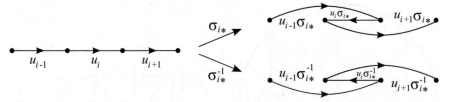

Fig. 4.7 The action (4.20), with the free loops in Fig. 4.6 depicted as arcs between punctures

fore and after passing to free homotopy. Symbolically, in free homotopy, two loops are equal if they are conjugate in $\pi_1(S, x_0)$, so that $e_i u_i^{\pm 1} e_i^{-1}$ is the same as $u_i^{\pm 1}$. We can thus rewrite the action (4.18)–(4.19) in the compact form

$$u_j \sigma_{i*}^{\pm 1} = \begin{cases} u_{i-1} u_i, & j = i - 1; \\ u_i^{-1}, & j = i; \\ u_i u_{i+1}, & j = i + 1; \\ u_j, & \text{otherwise.} \end{cases} \tag{4.20}$$

This is now closed, in the sense that it only involves the u_i. We can regard the u_i again as elements of a free group F_{n-1}. The action gives a representation for B_n in terms of *outer automorphisms* $\mathrm{Out}(F_{n-1})$ of F_{n-1}, which are elements of $\mathrm{Aut}(F_{n-1})$ equivalent up to conjugacy. The action (4.20) has an elegant pictorial representation in terms of arcs between the punctures, shown in Fig. 4.7. The figure also makes it clear that σ_{i*}^{-1} has exactly the same action as σ_{i*}^{+1} when written in terms of free generators, so the representation is not very faithful.

4.5 The Burau Representation

As we saw in Sect. 4.3, the Artin representation on $\pi_1(D_n, x_0)$ is not easy to manipulate once braids become complicated. Abelianization turns everything into linear algebra, which is desirable since matrix multiplication is fast and easy, but does not help since we recover a simple representation for the symmetric group on n symbols in terms of adjacent permutations. Too much is lost in the process of Abelianizing. Now we look for a compromise: is there a way to obtain a matrix representation that is more faithful to the braid group, i.e., not as disappointingly simple as the symmetric group? Here we construct such a representation, due to Burau [18, 19, 30].

Figure 4.8a shows the disk D_3 with a generating set for its fundamental group. We have added three *branch cuts*: whenever a loop crosses such a cut, it now exists on a new copy (or *deck*) of the punctured disk. This collection of decks is called a *branched cover* of D_n, denoted $\widetilde{D}_n$. This construction is shown schematically in Fig. 4.8b. Note that the three cuts share the same new copy of the disk (unlike the *universal cover*, where each cut would get its own new copy). It is easy to see how

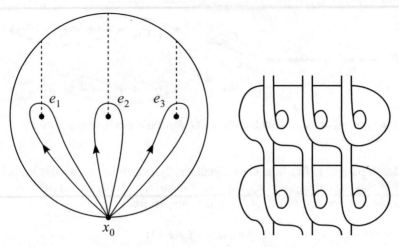

Fig. 4.8 (a) The disk with 3 punctures and a generating set for $\pi_1(D_3, x_0)$. The dashed lines are branch cuts. (b) Branched cover $\widetilde{D}_3$ corresponding to the cuts on the left. Two decks are shown, but the cover can be infinite in each direction, or the decks at each end can eventually join up

to generalize this construction to more punctures. We call this new space a $\mathbb{Z}$-*cover* of the disk D_n, since each deck can be indexed by a single number in $\mathbb{Z}$. (See the lectures by Boyland and Franks [25].)

When lifted to the $\mathbb{Z}$-cover, a loop e_i is no longer closed and becomes an arc. It joins the basepoint $\tilde{x}_0$ to another point $t\tilde{x}_0$, where t is a *deck transformation* that moves a point from one deck of the cover to the next one. In general, t^m with $m \in \mathbb{Z}$ moves us from the fundamental domain (the original disk) to the mth deck. We write $\tilde{e}_i$ for the unique *lift* of e_i to the $\mathbb{Z}$-cover. Then $t^m \tilde{e}_i$ is an oriented curve joining the basepoint in deck $m-1$ to the basepoint in deck m. In Fig. 4.9 (left), we represent the oriented curves and the different decks of the $\mathbb{Z}$-cover as a graph (the *1-skeleton* of the $\mathbb{Z}$-cover).

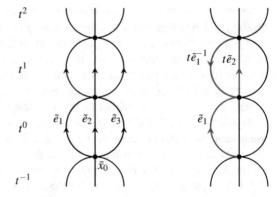

Fig. 4.9 Left: the 1-skeleton of the $\mathbb{Z}$-cover $\widetilde{D}_3$. Right: the action of σ_{1*} on $\tilde{e}_1$, as given by Eq. (4.22) with $i = 1$

Now consider the same action as (4.10) induced by σ_{i*} on the $\mathbb{Z}$-cover $\tilde{D}_n$. We have

$$\tilde{e}_i \, \sigma_{i*} = \tilde{e}_i (t\tilde{e}_{i+1})(t\tilde{e}_i^{-1}), \tag{4.21a}$$

$$\tilde{e}_{i+1} \, \sigma_{i*} = \tilde{e}_i, \tag{4.21b}$$

$$\tilde{e}_j \, \sigma_{i*} = \tilde{e}_j, \qquad j \neq i, i+1. \tag{4.21c}$$

Equation (4.21a) is depicted on the 1-skeleton in Fig. 4.9 (right) for $i = 1$ and $n = 3$. An easy way to see how this works is to remember that the curves are continuous, so curve $\tilde{e}_1$ must be followed by a curve that begins at $t\tilde{x}_0$, in this case $t\tilde{e}_2$. The final segment $t\tilde{e}_1^{-1}$ actually begins at $t^2\tilde{x}_0$ since it is the reverse of $t\tilde{e}_1$.

What have we gained by lifting to the $\mathbb{Z}$-cover? So far, not much. But the trick is that now we Abelianize (4.21). Writing again $\tilde{E}_i$ for the Abelianized version of $\tilde{e}_i$, we find Eq. (4.21a) becomes

$$\tilde{E}_i \, \sigma_{i*} = \tilde{E}_i + t\tilde{E}_{i+1} - t\tilde{E}_i = (1-t)\tilde{E}_i + t\tilde{E}_{i+1} \tag{4.22}$$

where following convention we now use addition and subtraction to write the Abelian composition law. Because of the parameter t, there is no cancellation in (4.22), and so the expression does not reduce to a simple permutation (unless we set $t = 1$). The representation of B_n we obtain is now in terms of an action on $H_1(\tilde{D}_n, \mathbb{Z})$, the first homology group of $\tilde{D}_n$ with integer coefficients (regarded as a module over the ring $\mathbb{Z}[t, t^{-1}]$). Elements of the homology group are linear combinations of terms of the form $t^m \tilde{E}_i$.

Using (4.22) and Eqs. (4.21b)–(4.21c) (which are unchanged by Abelianization), we can thus represent the action of σ_{i*} as a matrix, which gives the *Burau representation* of B_n [18, 19, 30]:

$$\sigma_i \mapsto I_{i-1} \oplus \begin{pmatrix} 1-t & t \\ 1 & 0 \end{pmatrix} \oplus I_{n-i-1}, \tag{4.23}$$

where I_m denotes an m by m identity matrix. The Burau representation maps B_n to $GL_n(\mathbb{Z}[t, t^{-1}])$, the general linear group with entries that are integer polynomials in t and t^{-1}.

The matrices of the Burau representation have an invariant vector

$$\tilde{U}_n = \tilde{E}_1 + t\tilde{E}_2 + \cdots + t^{n-1}\tilde{E}_n, \tag{4.24}$$

which corresponds to the lift of a loop around all the punctures, $e_1 e_2 \cdots e_n$. Such a loop is clearly invariant under any diffeomorphism induced by the σ_is. If we now change to a basis $\tilde{U}_i = \tilde{E}_i - \tilde{E}_{i+1}$, $1 \leq i \leq n-1$, and $\tilde{U}_n$ as above, the matrices of the representation block up, with a 1×1 block in the bottom-right slot. We may thus drop the last coordinate to obtain the *reduced Burau representation*, which maps B_n to $GL_{n-1}(\mathbb{Z}[t, t^{-1}])$:

$$[\sigma_i](t) = I_{i-2} \oplus \begin{pmatrix} 1 & t & 0 \\ 0 & -t & 0 \\ 0 & 1 & 1 \end{pmatrix} \oplus I_{n-i-2}. \tag{4.25}$$

Here for $i = 1$, we interpret I_{i-2} as deleting the first row and column of the 3×3 block, and for $i = n - 1$, we interpret I_{n-i-2} as deleting the last row and column of that block. The $-t$ entry in (4.25) is always in the diagonal (i,i) position. Thus, for $n = 3$, we have

$$[\sigma_1](t) = \begin{pmatrix} -t & 0 \\ 1 & 1 \end{pmatrix}, \qquad [\sigma_2](t) = \begin{pmatrix} 1 & t \\ 0 & -t \end{pmatrix}, \tag{4.26}$$

and for $n = 4$,

$$[\sigma_1](t) = \begin{pmatrix} -t & 0 & 0 \\ 1 & 1 & 0 \\ 0 & 0 & 1 \end{pmatrix}, \qquad [\sigma_2](t) = \begin{pmatrix} 1 & t & 0 \\ 0 & -t & 0 \\ 0 & 1 & 1 \end{pmatrix}, \qquad [\sigma_3](t) = \begin{pmatrix} 1 & 0 & 0 \\ 0 & 1 & t \\ 0 & 0 & -t \end{pmatrix}. \tag{4.27}$$

From now on when we refer to the Burau representation, we shall usually mean the reduced version.

Note that the homology basis $U_i = E_i - E_{i+1}$ corresponds to the homotopy loops $u_i = e_i e_{i+1}^{-1}$ of (4.14). In fact, we could have obtained the reduced Burau representation by starting from (4.18) and lifting to the $\mathbb{Z}$-cover. The e_i then cancel, and we recover the matrices (4.25).

What is the role of the parameter t? So far we have treated it simply as a symbol. But it can be useful to assign an actual complex value to t. For $t = 1$, we can see from (4.23) that the (unreduced) Burau representation just becomes a representation of the symmetric group, so that particular value should be avoided as it loses too much information about B_n. If we take t to be a root of unity with $t^M = 1$, then the $\mathbb{Z}$-cover becomes a finite cover of degree M (a $\mathbb{Z}_M$-cover).

For $t = -1$, we get a double cover ($t^2 = 1$). For D_3 in particular, observe that with $t = -1$ the matrices (4.26) are identical to the matrices T_1 and T_2 of the Dehn twists on the torus given by (2.10). This is because the double cover reverses the quotient of the torus by the hyperelliptic involution that we took in Chap. 2. Thus, for $t = -1$ and $n = 3$, the reduced Burau matrices generate $SL_2(\mathbb{Z})$. The case $t = -1$ is called the *integral Burau representation*. Other roots of unity for t have also been considered as special cases of the Burau representation [58, 59, 131].

Is the Burau representation faithful? Recall that the representation is faithful if the kernel of the map $B_n \to GL_{n-1}(\mathbb{Z}[t, t^{-1}])$ is trivial. This means that the only braid that maps to the identity matrix is the identity braid. It has long been known that the representation is faithful for $n \leq 3$. Then Moody [86] showed it is unfaithful for $n \geq 9$. Long and Paton [80] improved this to $n \geq 6$. Finally, Bigelow [13] showed that it is unfaithful for $n = 5$. This leaves the case $n = 4$, which is still open. However, Bigelow [14] showed that another matrix representation, the Lawrence–Krammer representation, is faithful for all n.

The lack of faithfulness of the Burau representation should not trouble us too much: it does not affect the central purpose for which we would like to use the Burau representation—to get lower bounds on the dilatation of braids. We shall see how to do this in Chap. 6.

4.6 Summary

In this chapter, we introduced the Artin braid group B_n for the disk with n punctures:

- Geometric braids are embedded in $\mathbb{R}^3$. Algebraic braids are their symbolic representation and are much more convenient to manipulate.
- Artin's representation of B_n is in terms of an action on the fundamental group π_1 (D_n), which is a free group.
- This representation is faithful but is unwieldy to compute with: it is hard to iterate since the length of the word in the free group can grow exponentially.
- We can Abelianize and instead get an action on the homology group $H_1(D_n, \mathbb{Z})$. However, we lose too much: we only retain information about the permutation induced by a braid.
- A middle ground is to lift Artin's action to a covering space $\widetilde{D}_n$ of D_n. After Abelianizing, we get the Burau representation of B_n. While not in general faithful, this representation is very easy to manipulate, since it is in terms of matrices with elements in $\mathbb{Z}[t, t^{-1}]$, where t is a parameter.

Our main use for the Burau representation will be in estimating the dilatation of a mapping class (Chap. 6). Some recent work of Yeung et al. [132] uses the Burau representation to detect the presence of invariant structures in flows. The Burau representation is surprisingly adept at this, and the algorithm is simpler and faster than the one used by Allshouse and Thiffeault [3], whose work was based on curves and Dynnikov coordinates (Chap. 8).

Chapter 5
The Thurston–Nielsen Classification

The first step in wisdom is to know the things themselves; this notion consists in having a true idea of the objects; objects are distinguished and known by classifying them methodically and giving them appropriate names. Therefore, classification and name-giving will be the foundation of our science.
—*Carl Linnaeus*, Systema Naturae

In Chap. 2, we gave a detailed description of the mapping class group of the torus, $\mathrm{MCG}(T^2) \approx \mathrm{SL}_2(\mathbb{Z})$. We encountered three main types of behaviors: finite-order ($|\tau| < 2$), Dehn twists ($|\tau| = 2$), and Anosov ($|\tau| > 2$). Two Dehn twists and their inverses can be composed to generate any element of the mapping class group. The most interesting case was the Anosov one since it led to rapid growth of curves under iteration, which could be exploited to analyze efficient taffy pullers and mixing devices (Chap. 3). In this chapter, we will see that these three types of behaviors carry over to more general surfaces than the torus, with some interesting complications.

5.1 Classification of Diffeomorphisms of a Surface

So far in the previous chapters, we have dealt with some basic surfaces: the torus T^2, the sphere with four punctures $S_{0,4}$, and the disk with n punctures D_n. We can unify the notation by writing $S_{g,n}^b$ for a surface of genus g with n punctures and b boundary components. Thus, $T^2 = S_{1,0}^0$ and $D_n = S_{0,n}^1$. We often omit b or n if either is zero. For instance, we write S_0 for a sphere, S_0^1 for D (disk with no punctures), and S_1 for T^2 (torus). Recall that the *Euler characteristic* $\chi(S_{g,n}^b)$ of a surface $S_{g,n}^b$ of genus g with n punctures and b boundary components is

$$\chi(S_{g,n}^b) = 2 - 2g - (b+n). \tag{5.1}$$

The following theorem was proved by Thurston [125], who built upon earlier work of Nielsen [68, 93]:

Theorem 5.1 (Thurston–Nielsen Classification). *Let $\phi \in \mathrm{Diff}^+(S)$, with S an orientable surface. Then $\phi \simeq \psi$, where ψ is a homeomorphism in the following three categories:*

© The Author(s), under exclusive license to Springer Nature Switzerland AG 2022
J.-L. Thiffeault, *Braids and Dynamics*, Frontiers in Applied Dynamical Systems:
Reviews and Tutorials 9, https://doi.org/10.1007/978-3-031-04790-9_5

1. Finite-order: *For some integer $k > 0$, $\psi^k \simeq$ id.*
2. Reducible: *ψ leaves invariant a disjoint union of essential simple closed curves, called* reducing *curves.*
3. Pseudo-Anosov: *ψ leaves invariant a transverse pair of measured singular folia-tions, $(\mathscr{F}_u, \mu_u)$ and $(\mathscr{F}_s, \mu_s)$, such that $\psi(\mathscr{F}_u, \mu_u) = (\mathscr{F}_u, \lambda \, \mu_u)$ and $\psi(\mathscr{F}_s, \mu_s) = (\mathscr{F}_s, \lambda^{-1} \mu_s)$, for dilatation $\lambda > 1$.*

Furthermore, if ψ is pseudo-Anosov it is not finite-order or reducible.

We say the mapping class of ϕ is finite-order (periodic), reducible, or pseudo-Anosov. The mapping ψ is sometimes called a *Thurston representative* or *model* of the mapping class of ϕ. It is a mapping that distills the essential behavior of ϕ, getting rid of all that can be ironed away by isotopy. We remark that in general ψ is only a homeomorphism since it is not differentiable at the singularities.

The theorem applies directly to the torus, and we have already encountered these three types of behaviors when discussing the classification of torus maps on pp. 16–19. The finite-order case was for $|\tau| < 1$ (elliptic); the reducible case was for $|\tau| = 2$ (parabolic); and the pseudo-Anosov case was for $|\tau| = 2$ (hyperbolic). For example, the reducing curves for T_1 consist of all curves parallel to $(0, 1)$.

The theorem contains a fair amount of terminology, so we will examine the three cases in detail and give an intuitive picture. We will describe here the first two cases and discuss the pseudo-Anosov case in its own section.

The finite-order or periodic case is the most straightforward: it simply says that eventually things return to their initial configuration, at least isotopically. It is important to note that the theorem does not imply that ϕ itself is periodic: it might still have very complicated behavior, but such behavior is not "forced" topologically. It can be isotoped to the identity after k iterations. It is actually always possible to take $\psi^k = $ id, with equality instead of isotopy.

The reducible case is characterized by a system of essential simple closed curves, $\Gamma = \{\Gamma_1, \ldots, \Gamma_m\}$ $(m \geq 1)$, which are pairwise-disjoint. The system of curves is invariant under ψ in the sense that $\psi(\Gamma) = \Gamma$, setwise, which means that the curves may be permuted by ψ. (Again, as in the finite-order case, Γ is not necessarily invariant under ϕ itself.)

For example, the curve $\pi\beta$ in Figs. 3.4 and 3.6 is invariant under $\overline{T}_1$ since the map only swaps the punctures p_1 and p_2 enclosed in that curve. In the reducible case, we reapply the theorem to each connected component of $S - \Gamma$, with ψ restricted to that component, possibly after taking a power of ψ to avoid permutations.

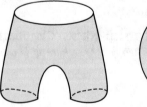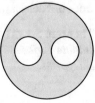

Fig. 5.1 A "pair-of-pants" (left) is the same as a disk with two boundary components removed (right). This surface has $\chi(S_0^3) = -1$

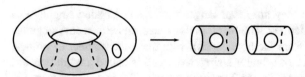

Fig. 5.2 The torus with two boundary components has $\chi(S_1^2) = -2$ and so can be cut by a maximum of two essential closed curves, yielding two pairs-of-pants with $\chi(S_0^3) = -1$

Another way to express the essential character of the curves in Γ is to require that each connected component of $S - \Gamma$ has negative Euler characteristic. The Euler characteristic of S places a bound on the number of connected components of $S - \Gamma$ and thus on the number of reducing curves. A maximal cut produces the "pants decomposition" of S: any surface with $\chi(S) < 0$ can be decomposed into $-\chi(S)$ pairs-of-pants (Fig. 5.1), which are spheres with three disks removed. Figure 5.2 shows the pants decomposition of a torus with two boundary components ($\chi(S_1^2) = -2$). Pairs-of-pants cannot be cut further without creating components of nonnegative Euler characteristic, which would imply that a Γ_i is nonessential.

Fig. 5.3 For the half-twist braid (4.6) acting on D_4, the two curves on the left and the curve on the right both form a system of reducing curves

The reducing curve system is not in general unique. For example, the half-twist braid defined in (4.6), which consists of rotating all the punctures by $180°$ as a set, leaves invariant both systems of curves in Fig. 5.3, for the case where the disk's outer boundary is fixed pointwise. A mapping class can be both reducible and finite-order, finite-order but not reducible, or reducible but not finite-order. But note that the theorem does not allow for a pseudo-Anosov mapping class to be also reducible or finite-order. See [45, p. 374] for some examples.

5.2 Pseudo-Anosov Maps

In the third case of Theorem 5.1, the map ψ was called pseudo-Anosov and was said to leave invariant a transverse pair of measured singular foliations, which requires some parsing. First, a foliation, for our purposes, is essentially a direction field (that is, a vector field with no arrows). It consists of infinite *leaves* that wind around the interior of S; a typical leaf is dense on S, except at boundaries, which are closed leaves. There are also distinguished semi-infinite leaves that end at *pronged singularities*, three examples of which are shown in Fig. 5.4. The presence of a finite number of singularities is what makes these "singular" foliations and is also what distinguishes Anosov from pseudo-Anosov maps. The two foliations, $\mathscr{F}_u$ and

$\mathscr{F}_s$, are *transverse*, in that they are never tangent, except at isolated singularities where transversality is ill-defined. The two foliations must share the same singularities in the interior (away from the boundaries), with the same number of prongs. A 2-pronged singularity is not actually a singularity and is called a *regular point*. Nevertheless, treating regular points as 2-pronged singularities is useful in some index formulas, such as (5.3) below. Pronged singularities can exist anywhere on the surface, except for a 1-pronged singularity that can only occur at a puncture or a boundary component.

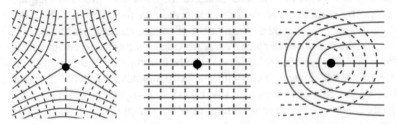

Fig. 5.4 Three types of pronged singularities of foliations. Left: 3-pronged singularity; middle: 2-pronged singularity; right: 1-pronged singularity. A 2-pronged singularity is actually not a singularity at all and is called a regular point. A 1-pronged singularity must coincide with a puncture

The measure μ on the foliations $\mathscr{F}$ is called a transverse measure. Given an arc α transverse to the foliation $\mathscr{F}$ (Fig. 5.5), the transverse measure $\mu(\alpha)$ is

$$\mu(\alpha) = \int_\alpha |\omega|, \tag{5.2}$$

where ω is a closed real-valued 1-form, which is locally defined. (See for example [45, p. 304].) The leaves of $\mathscr{F}$ correspond to the kernel of ω. If we write y for a coordinate perpendicular to the leaves, then we may take $\omega = c\,dy$. In some sense, μ measures the density of leaves intersecting an arc. The expression $\psi(\mathscr{F}_u, \mu_u) = (\mathscr{F}_u, \lambda\,\mu_u)$ in Theorem 5.1 means that the pseudo-Anosov map ψ sends leaves of $\mathscr{F}_u$ to leaves of $\mathscr{F}_u$ and increases the transverse measure μ_u by $\lambda > 1$. Conversely, the measure μ_s is decreased by λ^{-1}.

Boundaries add a few complications to this picture, which is why they are omitted in many treatments. However, boundaries are essential for applications, which involve rods and containers. The boundaries, if any, are special closed leaves of both foliations. At boundaries, the two foliations $\mathscr{F}_u$ and $\mathscr{F}_s$ need not have the same singular points, as depicted in Fig. 5.6. The foliations also need not be transverse to each other. See Exposé 11 of [46, 47] for a full discussion.

Figure 5.7 shows a very good approximation to the singular $\mathscr{F}_u$ that arises when stirring fluids. Here we have a very viscous fluid (a *Stokes flow*) in a circular container. The rods are treated as punctures, so the surface S is D_4. They are moved according to the braid $\sigma_1\sigma_2\sigma_3^{-1}$ (see Chap. 4). A curve is initially placed somewhere in the fluid and allowed to stretch and fold with the flow. Asymptotically, this curve will trace out $\mathscr{F}_u$, which is the pattern visible in Fig. 5.7. Note that $\mathscr{F}_u$ is only

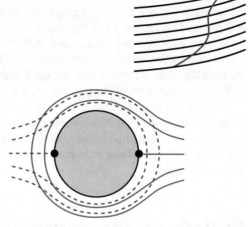

Fig. 5.5 An arc α transverse to a foliation: at no point is the arc tangent to the leaves of the foliation

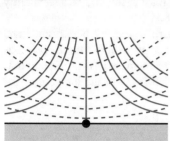

Fig. 5.6 Left: boundary singularity of $\mathscr{F}_u$. The solid lines denote leaves of the unstable foliations, and the dashed ones the stable foliation (which does *not* have a singularity in this picture). Right: a singularity of $\mathscr{F}_u$ and $\mathscr{F}_s$ around a rod, which can be regarded as a blow-up of the 1-pronged singularity in Fig. 5.4

partially traced out here, but if we waited longer, the picture would fill up all the way to the outer boundary. (This can, however, take a long time [61, 62, 123].) The line in Fig. 5.7 can also be thought of as one leaf of the foliation $\mathscr{F}_u$. We could in principle visualize $\mathscr{F}_s$ by running the experiment backward in time, which would resemble a flipped version of $\mathscr{F}_u$. For the Anosov case of the torus (Sect. 2.4.3), the loop in Fig. 2.5 can also be seen to be converging to an unstable foliation given by parallel lines of irrational slope, but in that case there are no singularities.

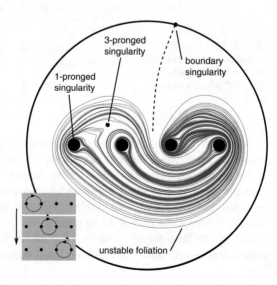

Fig. 5.7 A disk with four rods (punctures) following the braid $\sigma_1 \sigma_2 \sigma_3^{-1}$ (inset). A viscous fluid is simulated, with an idealized blob of dye filamenting to trace out the unstable foliation (after Thiffeault et al. [122])

The presence of singularities can be deduced visually in Fig. 5.7. Most obviously, around each rod is a 1-pronged singularity of the type shown in Fig. 5.4 (middle). A 3-pronged singularity as in Fig. 5.4 (left) is also apparent if we imagine that we fill in the leaves between the first and second rods: they must eventually meet at a 3-way intersection, which is the singularity. Finally, mentally filling in the leaves between the third and fourth rods, we see that there must eventually be a singularity on the outer boundary, of the type shown in Fig. 5.4 (right).

At each full period of the rod motion, the pattern in Fig. 5.7 is deformed and folds back upon itself. It is self-similar: it looks the same as it did on the previous iterate, except that the transverse density of leaves has increased by the dilatation λ. The folding of the pattern onto itself is exactly what leads to the increase of the transverse measure μ_u.

5.3 The Degree of the Dilatation

Thurston [125] showed that the dilatation λ is an *algebraic number*, that is, it is the root of a polynomial with integer coefficients. Moreover, the dilatation is the eigenvalue of an integer matrix, and so it is a root of a *monic* integer polynomial—a polynomial with unit leading coefficient. This means that λ is an *algebraic integer*.

The degree of an algebraic number is the smallest integer polynomial degree necessary to represent that number. For the torus, the dilatation always has degree 2 since we saw in Chap. 2 that it always arises from 2 by 2 matrices. We will see in this section that the degree of the dilatation for a pseudo-Anosov map on D_n is bounded.

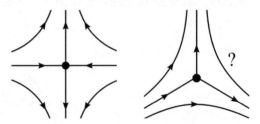

Fig. 5.8 We can define a consistent orientation around even-pronged singularities (left), but not around odd-pronged ones (right)

As we saw in Theorem 5.1, a pseudo-Anosov map has associated with it a pair of foliations, which share the same singularities. Sometimes we can assign a consistent global orientation to the foliation—that is, we can turn it into a vector field. In that case, the foliation is *orientable*. Figure 5.8 shows an obvious obstacle to orientability: if a singularity has an even number of prongs, then we can assign an orientation locally. However, we cannot do so when it has an odd number of prongs. Thus, a necessary condition for orientability is that all the singularities of a foliation have an even number of prongs. This condition is not sufficient: it is possible for a foliation to have only even singularities, and yet not be orientable due to the manner in which the singularities are arranged [83].

Recall that in Sect. 4.3, we discussed the action of a braid on first homology, $H_1(D_n, \mathbb{Z})$. Pseudo-Anosov maps that stabilize (leave invariant) orientable foliations have an astounding property: the largest eigenvalue of the matrix that gives their action on $H_1(S, \mathbb{Z})$ is exactly equal to the dilatation. (See for example [9, 17, 33, 45, 76, 77].) That is, unlike the situation in Sect. 4.3, the action on (Abelian) homology captures the dilatation perfectly. The reason is that the orientability of the foliation translates into an absence of cancellations when passing from $\pi_1(S)$ to $H_1(S, \mathbb{Z})$ by Abelianizing.

But as we saw at the end of Sect. 4.3, the action of a diffeomorphism on $H_1(D_n, \mathbb{Z})$ is always via permutation of the generators, which must lead to eigenvalues on the unit circle. Hence, it is never possible to infer that a map is pseudo-Anosov by looking at its action on $H_1(D_n, \mathbb{Z})$. (This is why we constructed the $\mathbb{Z}$-cover in Sect. 4.5.) This suggests that a foliation associated with a pseudo-Anosov map of D_n is *never* orientable, as is indeed the case.

Nevertheless, when faced with a nonorientable foliation, there is a standard construction to turn it into an orientable one. This construction, the *orientation double cover*, lies at the heart of many proofs, so it is worth learning. First, we give an example involving a nonorientable surface rather than a foliation since this is conceptually simpler. (This example is offered for flavor alone—we will not deal with nonorientable surfaces here, only nonorientable foliations.) Figure 5.9a shows the Möbius strip, the most classical nonorientable surface. In Fig. 5.9b, we cut the strip open, and in (c), we glue a new copy of the cut surface to the original along the exposed edges. The two copies are identified via rotation by 180° about the axis shown. The net result is an orientable surface that is a double cover of the Möbius strip. In fact here, this double cover is just a cylinder with two twists.

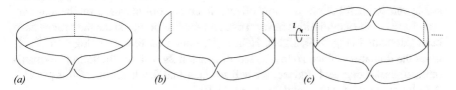

(a)   (b)   (c)

Fig. 5.9 (**a**) The Möbius strip, a nonorientable surface. (**b**) We make a cut along the strip and (**c**) glue two copies together at the open edges. The resulting surface is orientable, with the two halves identified by 180° rotation about the axis shown

Now we use a very similar trick to coerce a nonorientable foliation (on an orientable surface) into being orientable. Take the foliation of the pseudo-Anosov map on D_4 depicted in Fig. 5.7. There are five 1-pronged singularities: one at each rod, and one on the disk's boundary. There is also a 3-pronged singularity, which is not associated with a rod. All these singularities are odd-pronged, so the foliation is obviously nonorientable. There are a total of six odd-pronged singularities, which we depict schematically in Fig. 5.10 (left).

We then make branch cuts between pairs of singularities, which we show as dashed lines in Fig. 5.10a. We then cut along the branch cuts and shrink the outer boundary of D_4 to a fifth puncture (i.e., the reverse process to that of Fig. 3.5). In Fig. 5.10b, we have aligned the singularities and opened up the branch cuts. Finally, in Fig. 5.10c, we glue a second copy of the surface along the branch cuts, which gives a closed surface of genus two. This resulting surface is the orientation double cover $\widetilde{D}_4$ of D_4, branched at six points. The hyperelliptic involution ι maps the two halves to each other via rotation by $180°$, as shown in Fig. 5.10c.

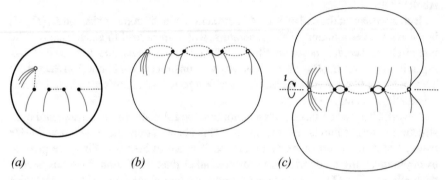

(a)   (b)   (c)

Fig. 5.10 (a) Disk D_4 with four 1-pronged singularities (punctures) and one 3-pronged singularity, corresponding to the foliation in Fig. 5.7. The dashed blue lines are branch cuts, and the prongs are denoted by red lines. (b) We collapse the disk's boundary to a puncture with a 1-pronged singularity and cut along the branch cuts. (c) Two copies of the middle picture are glued together to make the orientation double cover $\widetilde{D}_4$. The hyperelliptic involution ι maps the two copies to each other via rotation by $180°$ (deck transformation)

The upshot of the orientation double cover is that the lifts $\widetilde{\mathcal{F}}_u$ and $\widetilde{\mathcal{F}}_s$ of the unstable and stable foliations are now orientable on $\widetilde{D}_4$. For instance, on the new surface $\widetilde{D}_4 \approx S_2$, the singularities have an even number of prongs since the prongs have been duplicated (Fig. 5.10, right). Hence, the action on first homology $H_1(S_2, \mathbb{Z})$ gives the dilatation. Since $H_1(S_2, \mathbb{Z})$ has dimension $2g = 4$, the dilatation has maximum algebraic degree 4. Indeed, we will see later that the dilatation of this pseudo-Anosov map has exactly that degree (Eq. (7.10)).

In the general case of the punctured disk D_n, the Euler–Poincaré formula applied to a foliation $\mathcal{F}$ gives

$$\sum_{s \in \mathrm{sing}(\mathcal{F})} (2 - \mathrm{pr}(s)) = 2\chi(S_0) = 4, \tag{5.3}$$

where $\chi(S_0) = 2$ is the Euler characteristic of a sphere. Here $\mathrm{sing}(\mathcal{F})$ is the set of singularities of the foliation $\mathcal{F}$, and $\mathrm{pr}(s)$ is the number of prongs for the singularity s.[1] The formula (5.3) says that the number and type of singularities must add up

---

[1] In formula (5.3), we treat boundaries as punctures (singularities). Hence, boundaries such as in Fig. 5.6 are shrunk to pronged singularities as in Fig. 5.4.

in just the right way for the kind of surface they sit on: here we use a sphere, which is turned into a punctured disk D_n by puncturing and adding an outer boundary. For a torus, the right-hand side of (5.3) is zero, so that a foliation on an unpunctured torus need not have singularities. Formula (5.3) is also consistent with 2-pronged singularities with $\text{pr} = 2$ not really being singularities at all—they are regular points, as mentioned before.

We would like to maximize the number of odd-pronged singularities in the foliation, to obtain a bound on the degree of the dilatation. We denote by N_p the number of p-pronged singularities. The only way to get positive numbers on the left-hand side of (5.3) is to have a 1-pronged singularity with $\text{pr}(s) = 1$. The maximum number of such singularities on D_n is $N_1 = n + 1$: n on the punctures and 1 on the boundary. We want the remainder of the singularities to be 3-pronged, with each contributing $2 - 3 = -1$ to the left-hand side of (5.3). Hence, the maximum number N_3 of 3-pronged singularities satisfies

$$N_1 - N_3 = 4 \quad \Longleftrightarrow \quad N_3 = N_1 - 4 = n - 3. \tag{5.4}$$

Thus, a foliation can have up to $N_1 + N_3 = 2n - 2$ odd-pronged singularities. Passing to the orientation double cover will then give a surface $\widetilde{D}_n \approx S_g$ with $g = \frac{1}{2}(2n - 2) - 1 = n - 2$. (This is known as the Riemann–Hurwitz formula [89].) Finally, the dimension of $H_1(S_g, \mathbb{Z})$ is $2g = 2(n - 2)$ since each "handle" needs two homology generators. We conclude that λ has maximum algebraic degree $2n - 4$. Indeed, we already saw in Chap. 3 that for $n = 3$ the dilatation has degree $2 \times 3 - 4 = 2$ since a pseudo-Anosov map of D_3 lifts to the torus. In that case, the torus itself was the orientation double cover of D_3.

5.4 Summary

In this chapter, we explored the possible mapping class types on disks D_n, as given by the Thurston–Nielsen (TN) classification theorem.

- Finite-order (or periodic) maps are isotopic to the identity after some fixed number of iterates.
- Reducible maps leave invariant a system of disjoint curves. These curves may be permuted among themselves.
- Pseudo-Anosov maps stabilize (leave invariant) a pair of foliations, called stable and unstable. A transverse measure on each of these foliations is expanded or contracted by the dilatation, λ.
- It is not the original diffeomorphism ϕ that has these properties. Rather, the theorem says that ϕ is isotopic to a TN representative ψ that has those properties.
- For orientable foliations, the dilatation is easy to compute from the action on homology. But on D_n, no foliation is orientable.

- We can enforce an orientation by doubling up the surface—this is the orientation double cover.
- In practice, constructing the double cover requires knowledge of the types of singularities of a foliation, which are not known a priori.
- Nevertheless, the double-cover construction is very useful in proofs and for putting bounds on quantities such as the algebraic degree of the dilatation [33, 76, 77].

The TN classification applies to much broader classes of surfaces than punctured disks, but in this book we limit ourselves to them because they arise most often in applications.

We have already seen how to compute the dilatation for the case when the surface is D_3 (Chap. 3). In the next three chapters, we will see how to do this for disks with more punctures.

Chapter 6
Topological Entropy

Only entropy comes easy.
—Anton Chekhov

In Chap. 2, we introduced the dilatation λ of an Anosov mapping class of the torus. Then, in Chap. 5, we generalized the dilatation to the case of pseudo-Anosov diffeomorphisms of orientable surfaces. In both cases, $\log \lambda$ can be regarded as the growth rate of curves on a surface. To go back to our taffy puller examples of Chap. 1, $\log \lambda$ gives the rate at which taffy is stretched by the motion of the rods. In this chapter, we will relate $\log \lambda$ to the *topological entropy* of a diffeomorphism. We will also put some bounds on the topological entropy and hence on the dilatation.

6.1 Definition

Topological entropy was first introduced by Adler et al. [1] as a measure of dynamical complexity that only depends on topological features. We shall use the subsequent definition of Dinaburg [41, 42] and Bowen [21, 22] that requires a metric space but is independent of the metric and has a clearer interpretation. The topological entropy gives the exponential growth rate of the number of distinguishable orbits of a map, as we will now describe.

Let (S,d) be a compact[1] metric space and $\phi : S \to S$ a continuous map. For each integer $n \geq 0$, define a new metric on S by

$$d_n(x,y) = \max_{0 \leq k \leq n} d(\phi^k(x), \phi^k(y)). \tag{6.1}$$

Thus $d_n(x,y)$ is the maximum separation of the two points over orbits of length n. A subset E of S is (n,ε)-*separated* if each pair of distinct points x, y in E satisfies $d_n(x,y) \geq \varepsilon > 0$.

Notice that an (n,ε)-separated set consists of isolated points: if two points in E can be arbitrarily close, then we must take $\varepsilon = 0$. Smaller ε means that we can pack

---

[1] Though we work in the compact setting, note that the Dinaburg–Bowen definition generalizes to the noncompact case.

© The Author(s), under exclusive license to Springer Nature Switzerland AG 2022
J.-L. Thiffeault, *Braids and Dynamics*, Frontiers in Applied Dynamical Systems:
Reviews and Tutorials 9, https://doi.org/10.1007/978-3-031-04790-9_6

more points; larger n typically means that iterates come further apart, so we can also have more points.

Since E is discrete, let $N(n, \varepsilon)$ be the maximum cardinality of an (n, ε)-separated set, which is finite since S is compact. The quantity $N(n, \varepsilon)$ measures the maximum number of length-n distinguishable orbits of ϕ, assuming a "resolution" ε. The *topological entropy* is then

$$h(\phi) = \lim_{\varepsilon \to 0} \left\{ \limsup_{n \to \infty} \frac{1}{n} \log N(n, \varepsilon) \right\}, \tag{6.2}$$

where one can show that the limit exists. When S is compact and $\phi : S \to S$ is a homeomorphism, we have $h(\phi^n) = |n| h(\phi)$ [1, 21].

In the smooth context, we can relate the entropy to the growth of curves. Let $\phi : S \to S$ be a diffeomorphism of a smooth surface S. A smooth curve can be written as a smooth map $\alpha : \mathbb{I} \to S$, where $\mathbb{I} = [0, 1]$. We then define its *arc length* as

$$|\alpha| = \int |\alpha'(t)| \, dt, \tag{6.3}$$

where $|\alpha'(t)|$ is the length of the tangent in terms of some metric. A curve α has image $\phi^n(\alpha)$ after n iterates of ϕ. Then the topological entropy can be related to the maximum growth rate of smooth curves as [90, 91, 92, 133]

$$h(\phi) = \max_{\alpha} \limsup_{n \to \infty} \frac{1}{n} \log^+ |\phi^n(\alpha)|, \tag{6.4}$$

where $\log^+ x := \max(\log x, 0)$. For the case where ϕ is pseudo-Anosov, α can be any essential smooth curve, and we can drop the max. The relation (6.4) justifies our use of the growth rate of curves as a measure of topological entropy. We will now examine a similar idea: we look at the growth of elements of $\pi_1(S)$ under repeated action of the map ϕ_*.

6.2 Word Length Growth

A useful way to think of topological entropy comes from a bound due to Bowen [23]: he proved that for a homeomorphism ϕ of a surface S, $h(\phi)$ is bounded below by the *growth* of the action ϕ_* on $\pi_1(S)$. (We omit the basepoint for simplicity.) This measures how fast elements of $\pi_1(S)$ grow under repeated action of ϕ_*. We will first describe this in the context of an abstract group G.

Let $\mathcal{G} = \{e_1, \ldots, e_n\}$ be a set of generators for a finitely generated group G. We define the *reduced word length* $L_{\mathcal{G}}(\alpha)$ of an element $\alpha \in G$ as the minimum number of generators $e_i^{\pm 1}$ needed to write α. When G is a free group, this is easy to compute: all we have to do is to count generators after canceling adjacent inverses. A *reduced word* is a word $\alpha \in G$ written in terms of $L_{\mathcal{G}}(\alpha)$ generators. The reduced word length satisfies

$$L_{\mathscr{G}}(\alpha\beta) \le L_{\mathscr{G}}(\alpha) + L_{\mathscr{G}}(\beta), \qquad \alpha, \beta \in G. \tag{6.5}$$

For a group action $\phi_* : G \to G$, the *growth* of ϕ_* acting on G is

$$\mathrm{GR}(\phi_*) = \sup_{\alpha \in G} \limsup_{m \to \infty} (L_{\mathscr{G}}(\alpha\phi_*^m))^{1/m}. \tag{6.6}$$

Clearly, $\mathrm{GR}(\phi_*) \ge 1$. Since we are taking a supremum in G, we have

$$\mathrm{GR}(\phi_*) \ge \max_{1 \le i \le n} \limsup_{m \to \infty} (L_{\mathscr{G}}(e_i\phi_*^m))^{1/m}. \tag{6.7}$$

Now write a given α of reduced length r in reduced form $\alpha = e_{\mu_1}^{\varepsilon_1} e_{\mu_2}^{\varepsilon_2} \cdots e_{\mu_r}^{\varepsilon_r}$, with $\varepsilon_k = \pm 1$ and $1 \le \mu_k \le n$. Using the homomorphism property of ϕ_*^m and (6.5), we have

$$L_{\mathscr{G}}(\alpha\phi_*^m) = L_{\mathscr{G}}((e_{\mu_1}^{\varepsilon_1}\phi_*^m)(e_{\mu_2}^{\varepsilon_2}\phi_*^m) \cdots (e_{\mu_r}^{\varepsilon_r}\phi_*^m))$$
$$\le L_{\mathscr{G}}(e_{\mu_1}\phi_*^m) + L_{\mathscr{G}}(e_{\mu_2}\phi_*^m) + \cdots + L_{\mathscr{G}}(e_{\mu_r}\phi_*^m). \tag{6.8}$$

Before we continue, we prove a lemma (see [47, Lemma 10.6(i)]):

Lemma 6.1. *Let $(a_m)_{m \ge 1}$ and $(b_m)_{m \ge 1}$ be two sequences with a_m and b_m nonnegative. Then as $m \to \infty$,*

$$\limsup (a_m + b_m)^{1/m} = \max\left(\limsup a_m^{1/m}, \limsup b_m^{1/m}\right) = \limsup \max(a_m, b_m)^{1/m}.$$

Proof. Set $a = \limsup a_m^{1/m}$, $b = \limsup b_m^{1/m}$. We start by proving the first equality. Clearly,

$$\max(a, b) \le \limsup (a_m + b_m)^{1/m}.$$

Now we show the reverse inequality holds. If $c > \max(a, b)$, then we can find $m_0 \ge 1$ such that $m \ge m_0$ implies $a_m \le c^m$ and $b_m \le c^m$. Then for $m \ge m_0$,

$$(a_m + b_m)^{1/m} \le (2c^m)^{1/m}.$$

We thus have $\limsup (a_m + b_m)^{1/m} \le \limsup (2c^m)^{1/m} = c$.

For the second equality, we have that $\limsup \max(a_m, b_m)^{1/m} \ge \max(a, b)$. But also $\limsup \max(a_m, b_m)^{1/m} \le \limsup (a_m + b_m)^{1/m}$. $\qquad\square$

Taking the $\limsup$ on each side of (6.8) and using Lemma 6.1, we obtain

$$\limsup_{m \to \infty} (L_{\mathscr{G}}(\alpha\phi_*^m))^{1/m} \le \max_{1 \le i \le n} \limsup_{m \to \infty} (L_{\mathscr{G}}(e_i\phi_*^m))^{1/m}. \tag{6.9}$$

We conclude from (6.7) and (6.9) that we can rewrite (6.6) as

$$\mathrm{GR}(\phi_*) = \max_{1 \le i \le n} \limsup_{m \to \infty} (L_{\mathscr{G}}(e_i\phi_*^m))^{1/m}. \tag{6.10}$$

The second equality in Lemma 6.1 also implies that we can interchange the lim sup and max:

$$\mathrm{GR}(\phi_*) = \limsup_{m \to \infty} \max_{1 \le i \le n} (L_{\mathscr{G}}(e_i \phi_*^m))^{1/m}. \tag{6.11}$$

The form (6.11), which is in terms of generators rather than elements of G, makes it easy to show that the lim sup exists. Let $c = \max_i L_{\mathscr{G}}(e_i \phi_*)$. Then $L_{\mathscr{G}}(e_i \phi_*) \le c$, so that $L_{\mathscr{G}}(e_i \phi_*^m) \le c^m$ and $(L_{\mathscr{G}}(e_i \phi_*^m))^{1/m} \le c$. The argument of the lim sup in (6.10) is thus bounded above by c and below by 1 and so the lim sup must exist.

It is also easy to show that $\mathrm{GR}(\phi_*)$ does not depend on the set of generators $\mathscr{G}$. Indeed, if we took a different set of generators $\mathscr{G}' = \{e'_1, \ldots, e'_{n'}\}$, then clearly for any $\alpha \in G$,

$$L_{\mathscr{G}}(\alpha) \le \left(\max_{1 \le i \le n'} L_{\mathscr{G}}(e'_i) \right) L_{\mathscr{G}'}(\alpha). \tag{6.12}$$

The max in (6.12) is just a constant, and we can interchange $\mathscr{G}$ and $\mathscr{G}'$ in (6.12) to reverse the bound with a different constant. Positive multiplicative constants c do not affect the limit (6.10) since $\lim_{m \to \infty} c^{1/m} = 1$, so the growth is independent of the choice of generators for G.

Having reversed the order as in (6.11), we will now write $\mathrm{GR}(\phi_*)$ in terms of a matrix norm. The *occurrence matrix* $A(\phi_*)$ has entries $a_{ij}(\phi_*)$ equal to the number of occurrences of $e_j^{\pm 1}$ in the reduced word $e_i \phi_*$. We define the norm

$$\|A\| = \max_i \left(\sum_j |a_{ij}| \right). \tag{6.13}$$

Then $\max_i L_{\mathscr{G}}(e_i \phi_*^m) = \|A(\phi_*^m)\|$, and we can use the reversed definition (6.11) to write

$$\mathrm{GR}(\phi_*) = \limsup_{m \to \infty} \|A(\phi_*^m)\|^{1/m}. \tag{6.14}$$

Bowen [23], building on a theorem of Manning [82], showed that the topological entropy of a map ϕ is bounded from below by the growth of $\phi_* : \pi_1(S) \to \pi_1(S)$:

$$h(\phi) \ge \log \mathrm{GR}(\phi_*). \tag{6.15}$$

In fact for pseudo-Anosov maps of the type, we deal with here (see Chap. 5) this bound is an equality. The difficulty with Bowen's bound (6.15) is that the growth on $\pi_1(S)$ given by (6.10) is rather hard to compute, except in some simple cases. Many of the techniques we will introduce here and in the following chapters are essentially different methods of computing or estimating $\mathrm{GR}(\phi_*)$. In the next section, we describe a simple manner of getting an estimate for the case where S is a punctured disk.

6.3 The Burau Estimate for the Dilatation

For the case where $S = D_n$, the disk with n punctures, one of the most direct methods of estimating $h(\phi)$ is not an exact calculation, but rather to derive another bound. The treatment we give here is due to Kolev [75] who directly proved a theorem that is also a corollary of a result of Fried [57].

Recall that the mapping class group of D_n is isomorphic to the braid group B_n ((4.7) and (4.8)). Given a diffeomorphism $\phi : D_n \to D_n$, we denote by $\mathrm{br}(\phi) \in B_n$ the braid labeling the mapping class of ϕ. We will show the bound [75]

$$GR(\phi_*) \geq \sup_{|t|=1} \mathrm{spr}[\mathrm{br}(\phi)](t), \qquad (6.16)$$

where $[\mathrm{br}(\phi)](t)$ is the reduced Burau representation of $\mathrm{br}(\phi)$ from Sect. 4.5, Eq. (4.25), with parameter $t \in C$, and $\mathrm{spr}(M)$ is the *spectral radius* of the matrix M—the largest magnitude of its eigenvalues. The logarithm of the right-hand side of (6.16) is called the *Burau estimate* for the topological entropy of the braid $\mathrm{br}(\phi)$.

Now we introduce the *group ring* with integer coefficients $\mathbb{Z}F_n$ of the free group F_n. This contains formal linear combinations of elements of F_n with coefficients in $\mathbb{Z}$. For example, $e_1 e_3^{-1} + 2e_2^2 e_1 - e_4$ is an element of $\mathbb{Z}F_4$. For each j, we define the *free derivative* as an operator $\frac{\partial}{\partial e_j} : \mathbb{Z}F_n \to \mathbb{Z}F_n$ that satisfies [18, 55]

$$\frac{\partial}{\partial e_j}(e_i) = \delta_{ij}, \qquad\qquad 1 \leq i \leq n; \qquad (6.17a)$$

$$\frac{\partial}{\partial e_j}(\alpha_1 \alpha_2) = \frac{\partial}{\partial e_j}(\alpha_1) + \alpha_1 \frac{\partial}{\partial e_j}(\alpha_2), \qquad \alpha_1, \alpha_2 \in F_n; \qquad (6.17b)$$

$$\frac{\partial}{\partial e_j}(\ddot{\alpha}_1 + \ddot{\alpha}_2) = \frac{\partial}{\partial e_j}(\ddot{\alpha}_1) + \frac{\partial}{\partial e_j}(\ddot{\alpha}_2), \qquad \ddot{\alpha}_1, \ddot{\alpha}_2 \in \mathbb{Z}F_n. \qquad (6.17c)$$

From (6.17c) with $\ddot{\alpha}_1 = \ddot{\alpha}_2 = 0$, we get $\frac{\partial}{\partial e_j}(0) = 0$. From (6.17b) with $\alpha_1 = \mathrm{id}$, we get $\frac{\partial}{\partial e_j}(\mathrm{id}) = 0$. By setting $\alpha_1 = \alpha^{-1}$, $\alpha_2 = \alpha$ in (6.17b), we have

$$\frac{\partial}{\partial e_j}(\alpha^{-1}) = -\alpha^{-1}\frac{\partial}{\partial e_j}(\alpha), \qquad \alpha \in F_n. \qquad (6.18)$$

To get some practice, here are some basic examples:

$$\frac{\partial}{\partial e_j}(e_1 e_2) = \frac{\partial}{\partial e_j}(e_1) + e_1 \frac{\partial}{\partial e_j}(e_2) = \delta_{j1} + e_1 \delta_{j2}; \qquad (6.19)$$

$$\frac{\partial}{\partial e_j}(e_1 e_2^{-1}) = \frac{\partial}{\partial e_j}(e_1) + e_1 \frac{\partial}{\partial e_j}(e_2^{-1}) = \delta_{j1} - e_1 e_2^{-1}\delta_{j2}; \qquad (6.20)$$

$$\frac{\partial}{\partial e_j}(e_1^2) = \frac{\partial}{\partial e_j}(e_1) + e_1\frac{\partial}{\partial e_j}(e_1) = \delta_{j1} + e_1\delta_{j1}. \tag{6.21}$$

In fact, we can write down a general formula for $\frac{\partial}{\partial e_j}$ acting on a word in F_n. Let

$$\alpha = e_{\mu_1}^{\varepsilon_1}e_{\mu_2}^{\varepsilon_2}\cdots e_{\mu_r}^{\varepsilon_r} \in F_n, \qquad \varepsilon_k = \pm 1, \quad 1 \le \mu_k \le n, \tag{6.22}$$

be an arbitrary reduced word. Then

$$\frac{\partial}{\partial e_j}(\alpha) = \sum_{k=1}^{r} \varepsilon_k\delta_{\mu_k,j}e_{\mu_1}^{\varepsilon_1}e_{\mu_2}^{\varepsilon_2}\cdots e_{\mu_{k-1}}^{\varepsilon_{k-1}}e_{\mu_k}^{(\varepsilon_k-1)/2}. \tag{6.23}$$

We show this by induction. First, for $r = 1$, we have from (6.17a) and (6.18),

$$\frac{\partial}{\partial e_j}(e_{\mu_1}^{\varepsilon_1}) = \begin{cases} \delta_{\mu_1,j}, & \varepsilon_1 = 1; \\ -e_{\mu_1}^{-1}\delta_{\mu_1,j}, & \varepsilon_1 = -1, \end{cases} \tag{6.24}$$

which can be combined into one formula,

$$\frac{\partial}{\partial e_j}(e_{\mu_1}^{\varepsilon_1}) = \varepsilon_1\delta_{\mu_1,j}e_{\mu_1}^{(\varepsilon_1-1)/2}. \tag{6.25}$$

This obviously satisfies (6.23) with $r = 1$. Now assume the induction hypothesis (6.23) and consider a general word of length $r+1$. We use (6.17b) to expand the free derivative of the product:

$$\frac{\partial}{\partial e_j}(\alpha e_{\mu_{r+1}}^{\varepsilon_{r+1}}) = \frac{\partial}{\partial e_j}(\alpha) + \alpha\frac{\partial}{\partial e_j}(e_{\mu_{r+1}}^{\varepsilon_{r+1}})$$

$$= \left\{\sum_{k=1}^{r} \varepsilon_k\delta_{\mu_k,j}e_{\mu_1}^{\varepsilon_1}e_{\mu_2}^{\varepsilon_2}\cdots e_{\mu_{k-1}}^{\varepsilon_{k-1}}e_{\mu_k}^{(\varepsilon_k-1)/2}\right\} + \varepsilon_{r+1}\delta_{\mu_{r+1},j}\alpha e_{\mu_{r+1}}^{(\varepsilon_{r+1}-1)/2}$$

$$= \sum_{k=1}^{r+1} \varepsilon_k\delta_{\mu_k,j}e_{\mu_1}^{\varepsilon_1}e_{\mu_2}^{\varepsilon_2}\cdots e_{\mu_{k-1}}^{\varepsilon_{k-1}}e_{\mu_k}^{(\varepsilon_k-1)/2}.$$

This proves formula (6.23).

Remark 6.1. The sum in (6.23) is over the length of the reduced word α; the presence of the Kronecker symbol $\delta_{\mu_k,j}$ implies that the number of nonvanishing monomials in $\frac{\partial}{\partial e_j}(\alpha)$ will be equal to the number of occurrences of $e_j^{\pm 1}$. (It is easy to show that the individual monomials in (6.23) cannot combine or cancel if α is in reduced form.)

Now that we have defined the free derivative, and we need one more building block before we can use it for something interesting. Let $\langle t \rangle$ be the infinite cyclic group generated by t, and define the map

$$\varphi : F_n \to \langle t \rangle, \qquad \varphi : e_i \mapsto t, \qquad 1 \le i \le n. \tag{6.26}$$

This map just sums the powers of the generators in a given free word. For example, $\varphi(e_1^2 e_2 e_3^{-1}) = t^{2+1-1} = t^2$. The map φ extends to a homomorphism $\mathbb{Z}F_n \to \mathbb{Z}\langle t \rangle = \mathbb{Z}[t, t^{-1}]$. For example, $\varphi(3e_1^2 e_2 e_3^{-1} + 2e_2^{-1}) = 3t^2 + 2t^{-1}$.

Now comes the payoff: combining the free derivative $\frac{\partial}{\partial e_j}$ with the homomorphism φ allows us to write the entries of the matrices $B(\gamma, t) = (b_{ij}(\gamma, t))$ of the Burau representation of a braid γ as

$$b_{ij}(\gamma) = \varphi\left(\frac{\partial}{\partial e_j}(e_i \gamma_*)\right). \tag{6.27}$$

For instance, from Eq. (4.10a),

$$
\begin{aligned}
b_{ij}(\sigma_i) = \varphi\left(\frac{\partial}{\partial e_j}(e_i \sigma_{i*})\right) &= \varphi\left(\frac{\partial}{\partial e_j}(e_i) + e_i \frac{\partial}{\partial e_j}(e_{i+1} e_i^{-1})\right) \\
&= \varphi\left(\delta_{ij} + \delta_{j,i+1} e_i - e_i e_{i+1} e_i^{-1} \delta_{ij}\right) \\
&= (1-t)\delta_{ij} + t\, \delta_{i+1,j}.
\end{aligned}
$$

This corresponds to the first row of the 2 by 2 block in (4.23). The second row comes from

$$b_{i+1,j}(\sigma_i) = \varphi\left(\frac{\partial}{\partial e_j}(e_{i+1} \sigma_{i*})\right) = \varphi\left(\frac{\partial}{\partial e_j}(e_i)\right) = \delta_{ij}. \tag{6.28}$$

To prove directly that this is indeed a representation that requires deriving the "chain rule" for the free derivative (see Birman [18, p. 116]), but since the matrices correspond to our earlier Burau representation, we shall not need to do this.

Let us now apply φ to our general formula (6.23) for $\frac{\partial}{\partial e_j}(\alpha)$:

$$
\begin{aligned}
\varphi\left(\frac{\partial}{\partial e_j}(\alpha)\right) &= \sum_{k=1}^{r} \varepsilon_k \delta_{\mu_k, j} \varphi\left(e_{\mu_1}^{\varepsilon_1} e_{\mu_2}^{\varepsilon_2} \cdots e_{\mu_{k-1}}^{\varepsilon_{k-1}} e_{\mu_k}^{(\varepsilon_k - 1)/2}\right) \\
&= \sum_{k=1}^{r} \varepsilon_k \delta_{\mu_k, j} t^{\varepsilon_1 + \varepsilon_2 + \cdots + \varepsilon_{k-1} + (\varepsilon_k - 1)/2}.
\end{aligned}
$$

This is an element of $\mathbb{Z}[t, t^{-1}]$, so we write it as

$$\varphi\left(\frac{\partial}{\partial e_j}(\alpha)\right) = \sum_{|k| \leq r} c_k t^k, \tag{6.29}$$

for some set of coefficients c_k. Recall from Remark 6.1 that $\frac{\partial}{\partial e_j}(\alpha)$ has as many monomials as the number of occurrences of $e_j^{\pm 1}$ in α. Some of these monomials might give the same power of t in (6.29) and combine, so we have that the number of occurrences of $e_j^{\pm 1}$ is greater than $\sum |c_k|$. With $t \in C$, $|t| = 1$, we obtain the chain of inequalities [2]

$$a_{ij}(\phi_*) = \{\# \text{ of occurrences of } e_j^{\pm 1} \text{ in } e_i \phi_*\}$$

$$= \{\# \text{ of monomials in } \varphi\left(\frac{\partial}{\partial e_j}(e_i \phi_*)\right)\}$$

$$\geq \sum |c_k|$$

$$\geq \left|\sum c_k t^k\right| = \left|\varphi\left(\frac{\partial}{\partial e_j}(e_i \phi_*)\right)\right| = |b_{ij}(\phi,t)|,$$

where $b_{ij}(\phi,t)$ gives the entries of the Burau representation (6.27). This gives us an inequality on the occurrence matrix A and the Burau representation:

$$\|A(\phi_*^m)\| \geq \|B(\phi^m,t)\| = \|B^m(\phi,t)\|. \tag{6.30}$$

We can then use this in (6.14) to obtain finally

$$\mathrm{GR}(\phi_*) \geq \sup_{|t|=1} \limsup_{m \to \infty} \|B^m(\phi,t)\|^{1/m} = \sup_{|t|=1} \mathrm{spr}\, B(\phi,t). \tag{6.31}$$

This bound clearly still holds if we use the reduced Burau matrices (4.25) instead of the full Burau matrices since they differ only by a unit eigenvalue. Hence, the bound (6.16) follows immediately from (6.31).

Band and Boyland [9] have shown that the Burau estimate can be exact only for $t = -1$. That is, if the Burau estimate is equal to the topological entropy, then equality must occur only at $t = -1$. (See discussion of the double cover and $t = -1$ at the end of Sect. 4.5.) When the Burau estimate is not sharp, however, the best bound often occurs at other values of t (Fig. 6.1).

For braids in B_3, we have from (4.26)

$$[\sigma_1](-1) = \begin{pmatrix} 1 & 0 \\ 1 & 1 \end{pmatrix}, \qquad [\sigma_2](-1) = \begin{pmatrix} 1 & -1 \\ 0 & 1 \end{pmatrix}. \tag{6.32}$$

These are the same as the matrices $\overline{T}_1$ and $\overline{T}_2$ given in Eq. (1.1) and used in Sect. 3.3 to describe the action of Dehn twists on closed curves. Thus, in the case of B_3, the Burau estimate with $t = -1$ always gives the exact topological entropy for any braid.

For more than $n = 3$ strings, the Burau estimate is usually not sharp, and we must choose t to maximize the bound. For example, take the braid $\gamma = \sigma_1 \sigma_2 \sigma_3^{-1} \in B_4$. Its reduced Burau representation is

$$[\gamma](t) = [\sigma_1](t)[\sigma_2](t)[\sigma_3]^{-1}(t) = \begin{pmatrix} -t & t^2 & -t^2 \\ -1 & 0 & 0 \\ 0 & -1 & 1-t^{-1} \end{pmatrix}. \tag{6.33}$$

---

[2] Biryukov [20] relaxes the condition $|t| = 1$, but his bound is much harder to compute in practice.

This matrix has characteristic polynomial

$$p(x) = x^3 + (t + t^{-1} - 1)x^2 + t(t + t^{-1} - 1)x + t. \tag{6.34}$$

Now we find the optimal lower bound by writing $t = e^{i\theta}$ and varying theta between $-\pi$ and π. Figure 6.1 shows the dependence of the lower bound on θ: the optimal bound ≈ 2.174 is obtained for $\theta \approx \pm 2.496$. This is fairly close to the true value: in Sect. 7.2, we will use more powerful tools to find the exact dilatation for this braid, which is approximately equal to 2.3 (see Eq. (7.11)). In general, though, the Burau estimate can be poor, and in particular, it can sometimes be 1 (trivial) even for braids with positive entropy.

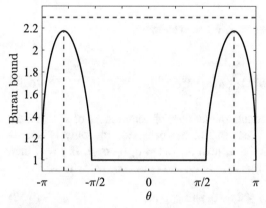

Fig. 6.1 The Burau lower bound on the dilatation of the braid $\sigma_1 \sigma_2 \sigma_3^{-1}$ with $t = e^{i\theta}$. The red dashed lines shows the Burau estimate (optimal lower bound) ≈ 2.174 for $\theta \approx \pm 2.496$. The blue dashed line is the actual dilatation ≈ 2.3 (see Eq. (7.11))

In light of the discussion of faithfulness on p. 40, note that any nontrivial non-central normal subgroup of B_n contains a pseudo-Anosov element [79, Lemma 2.5]. Hence, the unfaithfulness of the Burau representation for $n > 4$ necessarily implies that the Burau estimate is trivial for some pseudo-Anosov maps since the kernel of the representation is a normal subgroup. However, the questions of faithfulness and sharpness of the Burau bound do not seem to be closely related. For instance, the Lawrence–Krammer representation of B_n is faithful and can be used to obtain a lower bound on the topological entropy in a manner analogous to the Burau representation. In practice, there are many examples where this bound is not sharp, and it can be arbitrarily poor for some braids, despite the faithfulness of the representation.

6.4 An Upper Bound

Recall from Sect. 6.2, the occurrence matrix $A(\phi_*)$ has entries $a_{ij}(\phi_*)$ giving the number of occurrences of $e_j^{\pm 1}$ in the reduced word $e_i\phi_*$. From the action (4.10), we have

$$A(\sigma_{\ell *}) = I_{\ell-1} \oplus \begin{pmatrix} 2 & 1 \\ 1 & 0 \end{pmatrix} \oplus I_{n-\ell-1}, \tag{6.35}$$

and from (4.11),

$$A(\sigma_{\ell *}^{-1}) = I_{\ell-1} \oplus \begin{pmatrix} 0 & 1 \\ 1 & 2 \end{pmatrix} \oplus I_{n-\ell-1}. \tag{6.36}$$

We write $A \leq B$ if $a_{ij} \leq b_{ij}$ for every i, j. We then show

Lemma 6.2.

$$A(\phi_* \sigma_{\ell *}^{\pm 1}) \leq A(\phi_*)A(\sigma_{\ell *}^{\pm 1}). \tag{6.37}$$

Proof. The matrix entry $a_{ij}(\phi_*)$ counts the number of occurrences of $e_j^{\pm 1}$ in $e_i\phi_*$. Act with $\sigma_{\ell *}$ to make $e_i\phi_*\sigma_{\ell *}$. From (4.10), the generator $\sigma_{\ell *}$ only affects e_ℓ and $e_{\ell+1}$. If j is not equal to ℓ or $\ell + 1$, then $a_{ij}(\phi_*) = a_{ij}(\phi_*\sigma_{\ell *})$. If $j = \ell$, then from (4.10) the generator $\sigma_{\ell *}$ changes $e_{\ell+1}$ to e_ℓ, and e_ℓ to $e_\ell e_{\ell+1} e_\ell^{-1}$. Thus,

$$a_{i\ell}(\phi_* \sigma_{\ell *}) \leq 2a_{i\ell}(\phi_*) + a_{i,\ell+1}(\phi_*). \tag{6.38}$$

This is an inequality since $e_i\phi_*\sigma_{\ell *}$ could have some cancellations. If $j = \ell + 1$, then again from (4.10) we have

$$a_{i,\ell+1}(\phi_* \sigma_{\ell *}) \leq a_{i,\ell}(\phi_*). \tag{6.39}$$

From the form (6.35), the inequalities can be combined as $A(\phi_*\sigma_{\ell *}) \leq A(\phi_*)A(\sigma_{\ell *})$ element-by-element, and (6.37) is satisfied. We can repeat the same argument for $\sigma_{\ell *}^{-1}$.   □

We have immediately:

Corollary 6.1. *For a diffeomorphism* $\phi : D_n \to D_n$ *with braid representative* $\mathrm{br}(\phi) = \sigma_{\mu_1}^{\pm 1} \cdots \sigma_{\mu_k}^{\pm 1}$,

$$A(\phi_*) \leq A(\sigma_{\mu_1 *}^{\pm 1}) \cdots A(\sigma_{\mu_k *}^{\pm 1}). \tag{6.40}$$

Proof. Use Lemma 6.2 repeatedly.

The corollary used with (6.14) and the fact that $A \leq B$ implies $\|A\| \leq \|B\|$ for non-negative matrices gives:

Theorem 6.1. *For a diffeomorphism* $\phi : D_n \to D_n$ *with braid representative* $\mathrm{br}(\phi) = \sigma_{\mu_1}^{\pm 1} \cdots \sigma_{\mu_k}^{\pm 1}$,

$$GR(\phi_*) \leq spr\{A(\sigma_{\mu_1 *}^{\pm 1}) \cdots A(\sigma_{\mu_k *}^{\pm 1})\}. \tag{6.41}$$

Since the Bowen bound (6.15) is an equality for pseudo-Anosov diffeomorphisms, the theorem gives an upper bound on $h(\phi)$.

For example, take the braid $\gamma = \sigma_1 \sigma_2^{-1} \sigma_3$. We have the product

$$A(\sigma_{1*})A(\sigma_{2*}^{-1})A(\sigma_{3*}) = \begin{pmatrix} 2 & 1 & 0 & 0 \\ 1 & 0 & 0 & 0 \\ 0 & 0 & 1 & 0 \\ 0 & 0 & 0 & 1 \end{pmatrix} \begin{pmatrix} 1 & 0 & 0 & 0 \\ 0 & 0 & 1 & 0 \\ 0 & 1 & 2 & 0 \\ 0 & 0 & 0 & 1 \end{pmatrix} \begin{pmatrix} 1 & 0 & 0 & 0 \\ 0 & 1 & 0 & 0 \\ 0 & 0 & 2 & 1 \\ 0 & 0 & 1 & 0 \end{pmatrix} = \begin{pmatrix} 2 & 0 & 2 & 1 \\ 1 & 0 & 0 & 0 \\ 0 & 1 & 4 & 2 \\ 0 & 0 & 1 & 0 \end{pmatrix}.$$

The spectral radius of this matrix is ≈ 4.6167, so the bound from Theorem 6.1 is $GR(\gamma_*) \lesssim 4.6167$. The exact value is ≈ 3.7321.

6.5 Summary

In this chapter, we introduced the topological entropy of a map. This can be defined for very general maps and spaces, but here we specialize to pseudo-Anosov maps on punctured disks:

- The entropy is an upper bound for the rate of growth of words in $\pi_1(S)$, under repeated action of the induced map ϕ_*.
- When S is a punctured disk, we can lower bound the growth on $\pi_1(D_n)$ using the spectral radius of the Burau representation of the word. This is a quick-and-easy bound to compute, as it only requires linear algebra.
- This bound is sometimes sharp but it is not in general. Some words have positive entropy, even though their Burau estimate is zero.
- We can also easily get an upper bound on the growth on $\pi_1(D_n)$. This was used by Finn and Thiffeault [50], Boyland and Harrington [26], and Lynch [81] to prove optimality of certain braids, in the sense that their entropy per generator is maximized.

The Burau estimate is easy to compute but often fails badly. For this reason, we will cover two more methods of computing dilatation or topological entropy in the next two chapters. The first, train tracks (Chap. 7), is powerful and exact but is difficult to carry out. The second, Dynnikov coordinates (Chap. 8), is an iterative approach that is very expedient.

Chapter 7
Train Tracks

Trains begin to exist only when they derail.

—*Georges Pérec*

As we saw in Chap. 5, a central feature of pseudo-Anosov maps is that they stabilize (leave invariant) a pair of foliations, known as stable and unstable. In fluid dynamics, the unstable foliation is the "mixing pattern" that appears upon repeated stirring—see Figs. 1.1, 1.2, and 5.7. For taffy pullers, we can regard the taffy itself as a kind of foliation. The transverse invariant measure defined on the foliations, in both cases, is related to the density of "folds" that have appeared in the pattern. In fluid mixing, the invariant pattern is a topological analogue of the "strange eigenmode" that determines the rate of mixing [61, 62, 63, 70, 99, 105, 113, 120, 122, 130]. In this chapter we will see how Thurston captured the essence of the foliation by collapsing it to a graph he called a *train track* [87, 98]. Train tracks allow us to carry out explicit computations involving pseudo-Anosov maps.[1]

This chapter will not be very rigorous: we will introduce train tracks in an intuitive manner and explain their properties using examples. See Bestvina and Handel [12], Penner and Harer [98] for the complete theory.

7.1 The Figure-Eight Stirring Device

Before giving a more technical description of a train track, we will first proceed by an intuitive example. Figure 7.1 shows the mixing pattern for the figure-eight rod motion discussed in Chap. 1 (Fig. 1.1). The foliation consists of infinite leaves that wind their way around the rods. However, the leaves do not wind in an arbitrary manner: for instance, a leaf that plunges from above between the second and third rods always turns to the left. This is a consequence of the direction of the rod motion, as shown in Fig. 1.1.

We can now imagine "squishing" the mixing pattern, in a smooth manner, onto a one-dimensional skeleton. We might obtain something like the graph (actually a

---

[1] Some of the examples in this chapter are heavily influenced by Toby Hall's unpublished notes [65]. See also his section in the review by Boyland [24].

© The Author(s), under exclusive license to Springer Nature Switzerland AG 2022
J.-L. Thiffeault, *Braids and Dynamics*, Frontiers in Applied Dynamical Systems:
Reviews and Tutorials 9, https://doi.org/10.1007/978-3-031-04790-9_7

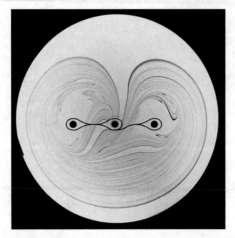

Fig. 7.1 The mixing pattern of the figure-eight rod motion in Fig. 1.1, with braid $\sigma_2^{-2}\sigma_1^2$. The dye pattern traces out the unstable invariant foliation. The corresponding train track is overlaid on the mixing pattern (From experiments by E. Gouillart and O. Dauchot.)

branched manifold) overlaid in Fig. 7.1. This graph is a train track corresponding to the unstable foliation (it is reproduced at the top of Fig. 7.3).

Fig. 7.2 A fibered neighbor-hood of a train track

Perhaps the reason for the name "train track" is more apparent now: Fig. 7.2 shows a so-called *fibered neighborhood* of a train track, where the stable foliation is depicted as crossties. In Thurston's notes [126, p. 204], he sketched a small train moving along the train track: a leaf of the foliation can wind around the fibered neighborhood, but it cannot double-back at the switches. It must travel along, obey-

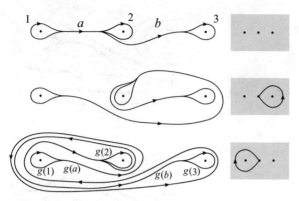

Fig. 7.3 A train track transformed by a figure-eight motion (braid $\sigma_2^{-2}\sigma_1^2$) of the central puncture. The right column shows the puncture's motion, and the left column gives the corresponding transformed graph

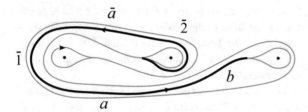

Fig. 7.4 The bottom frame of Fig. 7.3, with the image $g(b)$ of edge b emphasized. The image traces out $\bar{2}\bar{a}\bar{1}ab$ in terms of the *initial* edges

ing the switches as if it were a tiny train. Train tracks are a beautiful example of a naming convention that actually carries intuitive mathematical meaning.

How can we use a train track to compute something about a pseudo-Anosov map? In Fig. 7.3, we start at the top by labeling the different segments (called *edges*) of the train track and giving them an arbitrary orientation. We label the "loops" around the punctures by the numbers 1,2,3 and the two arcs between punctures by a and b. (The loops are also called *peripheral edges*, and edges such as a and b are called *main edges*.) We then move the central puncture in a figure-eight motion, dragging the train track as we go. Once we reached the end of the motion, we obtain a transformed train track, as in the bottom of Fig. 7.3. We write the image of edge a as $g(a)$, etc.

Now comes the tricky part. Observe that this final transformed train track can be smoothly "squished" to resemble the initial train track. By smoothly we mean that in doing this squishing, we do not need to create any kinks in the train track. This is the invariance property of the foliation reflected by the train track. The invariance allows us to write a map describing how the edges are transformed. Consider, for instance, the image of edge b shown in Fig. 7.4. We can see that in the final transformed train track, it now loops under puncture 2, goes to the left of puncture 1, and finally connects to puncture 2. We can rewrite this *edge path* in terms of the *initial* edges. Thus, the edges a, b, 1, 2, 3 are transformed by the map g as

$$a \mapsto a\bar{2}\bar{a}\bar{1}ab\bar{3}\bar{b}\bar{a}1a, \qquad b \mapsto \bar{2}\bar{a}\bar{1}ab, \qquad 1 \mapsto 1, \quad 2 \mapsto 2, \quad 3 \mapsto 3. \qquad (7.1)$$

Fig. 7.5 The edge path $b\bar{b}b$
(top) can be "pulled tight"
to b (bottom)

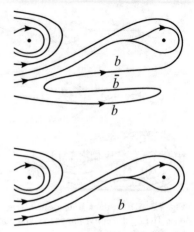

Here an overbar means that this particular segment of the transformed edge is oriented in the opposite direction to the initial edge. What we have essentially done is to turn the graphical description of the transformed train track into a symbolic representation. This symbolic representation is called a *train track map*. It is an automorphism of the free group generated by $\langle a,b,1,2,3 \rangle$.

Note that combinations such as $a\bar{a}$ or $b\bar{b}$ never occur in (7.1). These would cancel and correspond to the transformed train track somehow not being "pulled tight": there would be dangling bits of curve that can be shortened by isotopy (Fig. 7.5). However, it is not clear that when we iterate the map (7.1), such combinations will never arise. A train track map where cancellations never occur for any iterate is called *efficient*. In fact, the map g in (7.1) is efficient; we will return to this point later (Sect. 7.4).

A second iteration of the train track map g means applying the rule again to all the individual symbols on the right-hand side. We must also use the transformation rules for $\bar{a}$ and $\bar{b}$, which are obtained from (7.1) in the obvious way:

$$\bar{a} \mapsto \bar{a}\,\bar{1}ab3\bar{b}\bar{a}1a2\bar{a} \qquad \bar{b} \mapsto \bar{b}\bar{a}1a2. \tag{7.2}$$

The transformation rules for $\bar{1}$, $\bar{2}$, $\bar{3}$ are also trivial. Clearly, writing down even the second iterate of g is tedious. The reason is that as the train track map is iterated, the number of symbols grows exponentially, at a rate eventually given by the topological entropy, $\log\lambda$. This is because the number of symbols is directly related to the length of the curve entwined around the punctures. The rapid growth of the number of symbols makes the map g difficult to compute with.

However, we are in luck. Since the map g defined by (7.1) is efficient (though we have not yet shown this, see Sect. 7.4), no cancellations ever occur as we iterate g. Sequences such as $\bar{a}1a$ do occur, but the 1 in the middle acts as a buffer and prevents cancellation. Pictorially, this means that a piece of train track is folded over puncture 1. The lack of cancellations means that we can completely disregard the overbars, without fear of losing any symbols. Hence, we can Abelianize: we write the product operation in (7.1) as addition and drop the overbars:

$$a \mapsto a+2+a+1+a+b+3+b+a+1+a, \tag{7.3a}$$
$$b \mapsto 2+a+1+a+b. \tag{7.3b}$$

Since addition is commutative, after collecting terms, this simplifies to

$$a \mapsto 5a+2b +2(1)+(2)+(3), \tag{7.4a}$$
$$b \mapsto 2a+ b + (1)+(2). \tag{7.4b}$$

Note that there is momentarily some ambiguity between integer coefficients and the labels 1, 2, 3 for the edges around the punctures, so we have put parentheses around those edge labels.

We can now write (7.4) as a matrix transformation:

$$\begin{pmatrix} a \\ b \\ 1 \\ 2 \\ 3 \end{pmatrix} \mapsto \begin{pmatrix} 5 & 2 & 2 & 1 & 1 \\ 2 & 1 & 1 & 2 & 0 \\ 0 & 0 & 1 & 0 & 0 \\ 0 & 0 & 0 & 1 & 0 \\ 0 & 0 & 0 & 0 & 1 \end{pmatrix} \begin{pmatrix} a \\ b \\ 1 \\ 2 \\ 3 \end{pmatrix}. \tag{7.5}$$

This is the "Abelian" version of (7.1): it contains less information than the original train track map, since the order in which the symbols occur is forgotten. However, it does accurately reflect the number of edges of each type for each iterate. But the great advantage of (7.5) is that it is a linear matrix transformation, and we know how to compute the growth in the number of symbols in this case: they will grow asymptotically at a rate given by the spectral radius of the matrix, which is the magnitude of the largest eigenvalue.

In fact the spectral radius is simpler to find than it looks. The matrix in (7.5)—called a *transition matrix*—has a block structure. The lower-right block is the identity matrix (in general, it corresponds to the permutation of the punctures), so its largest eigenvalue has unit magnitude. The upper-left block is the matrix $\begin{pmatrix} 5 & 2 \\ 2 & 1 \end{pmatrix}$, which does have an eigenvalue with magnitude larger than unity. Hence, the spectral radius is "contained" entirely in the upper-left matrix block. Its characteristic polynomial is $x^2 - 6x + 1$, so the spectral radius is the largest root $\lambda = (1+\sqrt{2})^2 \approx 5.83$. The fact that the matrix in (7.5) blocks up is due to the edges 1, 2, 3 being only permuted: these peripheral edges do not contribute to the dilatation λ. However, the peripheral edges are crucial in preventing cancellations between the main edges.

Note that this is the same dilatation found in Sect. 3.5 for the three-rod taffy puller. This is not a coincidence: the two torus maps are conjugate to each other, since

$$\begin{pmatrix} 5 & 2 \\ 2 & 1 \end{pmatrix} = \begin{pmatrix} 1 & 1 \\ 0 & 1 \end{pmatrix} \begin{pmatrix} 3 & 4 \\ 2 & 3 \end{pmatrix} \begin{pmatrix} 1 & 1 \\ 0 & 1 \end{pmatrix}^{-1}. \tag{7.6}$$

Fig. 7.6 The initial train track (top) is acted on by the braid $\sigma_1\sigma_2\sigma_3^{-1}$, broken into two steps, from top to bottom: $\sigma_1\sigma_2$ and then σ_3^{-1}. The three-pronged singularity of the foliation (Fig. 5.4) corresponds to the shaded interior region with three cusps

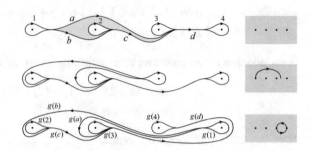

7.2 A Second Pseudo-Anosov Example

The example presented in the previous section is very simple, and it hides a lot of the difficulty in computing with train tracks. In addition, since the device only has three rods, it could have been analyzed with the methods of Chap. 3. (It is essentially the same as the taffy puller example in Sect. 3.5.)

A somewhat more complicated example is the motion depicted in Fig. 5.7, which is represented by the braid $\sigma_1\sigma_2\sigma_3^{-1}$. The main problem is to find the correct invariant train track. We postpone this problem to the next section, and with foresight use the train track in Fig. 7.6 (top). This train track should be compared to the mixing pattern in Fig. 5.7: the train track exhibits an interior region with three "cusps," which corresponds to a three-pronged singularity. The peripheral loops around each puncture have one cusp, which are one-pronged singularities [98]. Again, the structure of the train track reflects the fact that leaves wrapped around the first puncture can pass both above and below the second puncture, as can be seen in Fig. 5.7.

Figure 7.6 (bottom) shows the result of acting on the train track with $\sigma_1\sigma_2\sigma_3^{-1}$. We rewrite the image of the edges in terms of the initial edges to obtain the train track map g,

$$
\begin{aligned}
a &\mapsto d\bar{c}\bar{2}, & 1 &\mapsto 4, \\
b &\mapsto \bar{d}\bar{a}\bar{1}, & 2 &\mapsto 1, \\
c &\mapsto b, & 3 &\mapsto 2, \\
d &\mapsto cd\bar{4}\,\bar{d}, & 4 &\mapsto 3.
\end{aligned}
\tag{7.7}
$$

Again, we assume that this map is efficient. (We will show that this is so in Sect. 7.4.) We can then Abelianize: we rewrite multiplication in (7.7) as a sum and drop the overbars. The main edges (a, b, c, d) are then transformed according to the matrix

$$
N =
\begin{pmatrix}
0 & 0 & 1 & 1 \\
1 & 0 & 0 & 1 \\
0 & 1 & 0 & 0 \\
0 & 0 & 1 & 2
\end{pmatrix}.
\tag{7.8}
$$

How can we determine if this corresponds to a pseudo-Anosov map? Such a map should not contain any invariant curves and should not be finite-order. In terms of the matrix N, observe that

$$N^5 = \begin{pmatrix} 2 & 6 & 13 & 27 \\ 3 & 6 & 16 & 33 \\ 2 & 3 & 6 & 14 \\ 4 & 10 & 23 & 48 \end{pmatrix}, \tag{7.9}$$

that is, after 5 iterates, the matrix N^k is strictly positive.

Before we continue, we need a few facts from the theory of nonnegative matrices, i.e., matrices with all elements ≥ 0. A nonnegative matrix N is *irreducible* if for each i and j, there is a $k > 0$ such that $(N^k)_{ij} > 0$. An irreducible matrix is *primitive* if there exists a $k > 0$ such that $(N^k)_{ij} > 0$, for all i and j. (Primitive is stronger than irreducible, since the same k must work for all i, j.) For the train track map to correspond to a pseudo-Anosov map, the transformation matrix of main edges must be primitive. Clearly (7.9) shows that N is primitive.

A primitive matrix satisfies the following theorem [106].

Theorem 7.1 (Perron–Frobenius Theorem for Primitive Matrices). *Let N be a real primitive matrix. Then N has an eigenvalue λ such that*

1. *λ is real and strictly positive.*
2. *The corresponding eigenvector can be chosen real with nonnegative entries.*
3. *$\lambda > |\lambda'|$ for any eigenvalue $\lambda' \neq \lambda$.*
4. *λ is a simple root of the characteristic equation for N.*

Furthermore, an integer primitive matrix has $\lambda > 1$.

The primitivity of N ensures that any initial segment of train track will eventually propagate to the entire train track. This is the "topological mixing" property of pseudo-Anosov maps. The Perron–Frobenius theorem then guarantees that we can find a real positive dilatation λ from N. The normalized eigenvector associated with λ is a measure on the train track that reflects how dense the leaves of the foliations are along any segment of the train track (the transverse measure).

Thus, if we return to the matrix N in (7.8) above, it has characteristic polynomial

$$p(x) = x^4 - 2x^3 - 2x + 1. \tag{7.10}$$

This does indeed have a nondegenerate largest root, given by

$$\lambda = \tfrac{1}{2}(1 + \sqrt{3} + \sqrt{2}\, 3^{1/4}) = 2.29663026\ldots. \tag{7.11}$$

This is the dilatation of a pseudo-Anosov in the isotopy class labeled by $\sigma_1 \sigma_2 \sigma_3^{-1}$.

Fig. 7.7 The action of the braid (7.12) on a line diagram. The action is broken into four steps, from top to bottom: $\sigma_3\sigma_2\sigma_1$, σ_5, σ_3, and $\sigma_4\sigma_3\sigma_2\sigma_1$

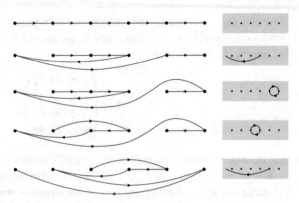

7.3 A Reducible Example

Our two examples so far were pseudo-Anosov braids. We assumed that we had the invariant train track at the outset, and we could verify that the braids were pseudo-Anosov. In practice, actually finding the invariant train track is the real challenge. The algorithm by Bestvina and Handel [12] is designed to do this: it starts with an initial guess at the train track and proceeds to modify and prune it to either find a *reduction* or achieve *efficiency*. The Bestvina–Handel algorithm actually amounts to a constructive proof of the classification theorem 5.1, since it terminates in a finite number of steps and yields the isotopy class. It is, however, an extremely difficult algorithm to write down and implement. A good place to learn more about it is in a section by Toby Hall contained in the review of Boyland [24]. Hall has also written a very useful C++ package that computes the train track map for any braid on D_n [64]; this code is used by the Matlab library braidlab [119].

Instead of explaining the Bestvina–Handel algorithm in detail, we will follow an example that illustrates some of the main ideas. We take the braid

$$\gamma = \sigma_3\sigma_2\sigma_1\sigma_5\sigma_3\sigma_4\sigma_3\sigma_2\sigma_1 . \tag{7.12}$$

Figure 7.7 shows the action of this braid on a *line diagram*—a simplified train track that consists of lines connecting the punctures. The line diagram is a reasonable place to start: interpreted physically, we wrap a piece of taffy over the entire puller and see what we get. The end result in Fig. 7.7 (bottom) does not suggest much about whether or not this braid labels a pseudo-Anosov class and if so what is the value of the dilatation.

The line diagram is missing the peripheral edges around each puncture. The bottom of Fig. 7.7 suggests that, to get a semblance of invariance, the initial train track should go below punctures 2–3, above or below puncture 4, and below puncture 5. We thus improve on our line diagram a little and redraw the initial train track as in Fig. 7.8 (top). Figure 7.8 also has main and peripheral edges labeled and gives the same train track map as we obtained from our line diagram. We call this train track graph G and define our train track map g:

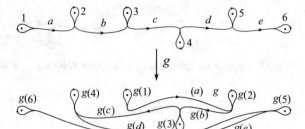

Fig. 7.8 The train track G (top) with map implied by the line diagram in Fig. 7.7 (bottom)

$$
\begin{aligned}
g(a) &= cd, & g(1) &= 3, \\
g(b) &= \bar{d}, & g(2) &= 5, \\
g(c) &= \bar{c}\bar{b}, & g(3) &= 4, \\
g(d) &= bc\bar{4}de, & g(4) &= 2, \\
g(e) &= \bar{e}\bar{d}4\bar{c}\bar{b}\bar{a}, & g(5) &= 6, \\
& & g(6) &= 1.
\end{aligned}
\tag{7.13}
$$

Is this a good train track map? More precisely, is it efficient in the sense that we defined before? Recall that an efficient map has no cancellations of the type $a\bar{a}$, as clearly (7.13) does not have, but also no cancellations in any iterates. But

$$
\begin{aligned}
g^2(d) &= g(bc\bar{4}de) \\
&= g(b)g(c)g(\bar{4})g(d)g(e) \\
&= (\bar{d})(\bar{c}\bar{b})(\bar{2})(bc\bar{4}de)(\bar{e}\bar{d}4\bar{c}\bar{b}\bar{a}) \\
&= \bar{d}\bar{c}\bar{b}\bar{2}\bar{a},
\end{aligned}
\tag{7.14}
$$

that is, there are cancellations in g^2 even though there were none in g itself. The map (7.13) is thus not efficient. A consequence of these cancellations is that if we Abelianize we will *overestimate* the dilatation, since iterates of the linearization will contain symbols that would have cancelled in the full map.

The source of the cancellation lies in that $g(d)$ contains de, and $g(\bar{d})$ and $g(e)$ start in the same way ($\bar{e}\bar{d}4\bar{c}\bar{b}$). Hence, to eliminate the cancellation, we modify the train track as follows: we identify $\bar{d}$ with the initial segment of e which has image $\bar{e}\bar{d}4\bar{c}\bar{b}$ and denote the remainder of edge e as e'; symbolically,

$$
e = \bar{d}e' \quad \Longleftrightarrow \quad e' = de.
\tag{7.15}
$$

This gives the modified train track graph G' shown in Fig. 7.9. This process is called *folding* two edges.

What has this folding achieved? The new train track map induced by γ is g' and can be obtained algebraically from g:

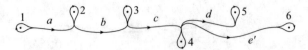

Fig. 7.9 The train track graph G', obtained from G by writing $e' = de$

$$g'(a) = cd,$$
$$g'(b) = \bar{d},$$
$$g'(c) = \bar{c}\bar{b}, \qquad (7.16)$$
$$g'(d) = bc\bar{4}de = bc\bar{4}\slashed{d}(\slashed{\bar{d}}e') = bc\bar{4}e',$$
$$g'(e') = g(de) = \bar{a}.$$

Instead of $(g')^2$, folding has pushed the cancellation to g', where we can explicitly cross it out. The resulting map g' is simpler than g, in a sense that can be made precise: upon Abelianizing, it has a smaller dilatation.

But we are not done: the map g' still has cancellations, since

$$(g')^2(a) = g'(c)g'(d) = (\slashed{\bar{c}}\bar{b})(\slashed{b}\bar{c}\bar{4}e') = \bar{4}e'. \qquad (7.17)$$

So the map g' is still not efficient. The problem this time is that $g'(a)$ contains cd, and $g'(\bar{c})$ and $g'(d)$ both begin with bc. The solution is to fold c and d:

$$d' = cd. \qquad (7.18)$$

The train track graph obtained from this folding is shown in Fig. 7.10. We update

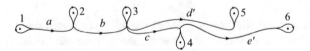

Fig. 7.10 The train track graph G'', obtained from G' by writing $d' = cd$

our graph map to g'':

$$g''(a) = cd = c\bar{c}d' = d',$$
$$g''(b) = \bar{d} = \bar{d'}c,$$
$$g''(c) = \bar{c}\bar{b}, \qquad (7.19)$$
$$g''(d') = \bar{4}e',$$
$$g''(e') = \bar{a}.$$

Notice that g'' acts on $\{a, d', e'\}$ in the following way:

$$(7.20)$$

The edges $\{a, d', e'\}$ (together with their peripheral edges) form an *invariant subgraph* of the train track graph. This subgraph is shown in Fig. 7.11.

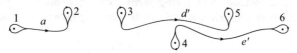

Fig. 7.11 Invariant subgraph of the train track G''

The existence of an invariant subgraph for the train track map immediately implies that the braid γ in (7.12) corresponds to a reducible isotopy class. The system of reducing curves can be read off the invariant subgraph and is shown in Fig. 7.12.

Fig. 7.12 System of reducing curves for γ

7.4 Finding Cancellations

So far in all our examples, we have sidestepped the issue of how to find cancellations (also called *backtracking*) or how to show that none occur for any iterate of the train track map.

Let us examine more closely the reducible example from the previous section. The cancellation that occurred in (7.14) was due to $g(\bar{d})$ and $g(e)$ starting with the same letters. Hence, the image of letters that start the same is potentially problematic. This suggests defining a map Dg, the *derivative map* of g, which consists of keeping only the first letter in each image of (7.13):

$$
\begin{aligned}
Dg(a) &= c, & Dg(\bar{a}) &= \bar{d}, \\
Dg(b) &= \bar{d}, & Dg(\bar{b}) &= d, \\
Dg(c) &= \bar{c}, & Dg(\bar{c}) &= b, \\
Dg(d) &= b, & Dg(\bar{d}) &= \bar{e}, \\
Dg(e) &= \bar{e}, & Dg(\bar{e}) &= a,
\end{aligned}
\tag{7.21}
$$

as well as $Dg(1) = 3$, etc. The action of Dg can be represented with the following diagram:

$$\bar{b} \longrightarrow d \qquad \bar{a} \qquad e$$

$$a \rightleftarrows c \longrightarrow \bar{c} \longrightarrow b \longrightarrow \bar{d} \longrightarrow \bar{e}$$

Two oriented edges E_1 and E_2 are in the same *gate* if they have the same initial point and $(Dg)^k(E_1) = (Dg)^k(E_2)$ for some $k \geq 0$. For example, $(Dg)^2(\bar{b}) = (Dg)^2(c) = b$, so $\bar{b}$ and c are in the same gate. The gates can be picked out from the diagram (7.4) by examining each junction and tracing backwards. There are three junctions (at b, d, and $\bar{e}$), which give four gates with more than one element:

$$\{\bar{b}, c\}, \quad \{\bar{c}, d\}, \quad \{\bar{a}, b\}, \quad \{\bar{d}, e\}. \tag{7.22}$$

From the gates, we create a list of *bad words*. For example, take the gate $\{\bar{b}, c\}$. The fact that these two letters are in the same gate means that for some future iterate their image begins with the same letter. Hence, the combination $\bar{\bar{b}}c = bc$ is bad, since the end of the image b can cancel with the start of the image of c. The bad words for a gate $\{E_1, E_2, \dots\}$ are thus formed by all combinations $\bar{E}_i E_j$, $i \neq j$. The list of bad words corresponding to the gates (7.22) is thus

$$\begin{array}{cccc} bc & cd & ab & de \\ \bar{c}\bar{b} & \bar{d}\bar{c} & \bar{b}\bar{a} & \bar{e}\bar{d}. \end{array} \tag{7.23}$$

We can see that many of these bad words appear in (7.13). We can thus expect cancellations, as indeed we found in Sect. 7.3. The goal of folding, and of several other operations involved in the Bestvina–Handel algorithm, is to eliminate these cancellations by rearranging the train track graph.

We can use the same kind of argument to show that the map (7.7) for the braid $\sigma_1 \sigma_2 \sigma_3^{-1}$ was indeed efficient, as we assumed. The derivative map is

$$\begin{array}{llll} a \mapsto \bar{d}, & \bar{a} \mapsto 2, & 1 \mapsto 4, & \bar{1} \mapsto \bar{4}, \\ b \mapsto \bar{d}, & \bar{b} \mapsto 1, & 2 \mapsto 1, & \bar{2} \mapsto \bar{1}, \\ c \mapsto b, & \bar{c} \mapsto \bar{b}, & 3 \mapsto 2, & \bar{3} \mapsto \bar{2}, \\ d \mapsto c, & \bar{d} \mapsto d, & 4 \mapsto 3, & \bar{4} \mapsto \bar{3}, \end{array} \tag{7.24}$$

with diagram

$$\bar{c} \longrightarrow \bar{b} \longrightarrow 1 \rightleftarrows 4 \longrightarrow 3 \longrightarrow 2$$

$$a \longrightarrow \bar{d} \rightleftarrows d \longrightarrow c \longrightarrow b$$

and a cyclic diagram for the peripheral edges $\bar{1}$–$\bar{4}$. The gates are thus

$$\{\bar{c}, 3\}, \quad \{\bar{b}, 2\}, \quad \{a, b\}, \tag{7.25}$$

which gives the list of bad words

$$c3 \; b2 \; \bar{a}b$$
$$3\bar{c} \; \bar{2}\bar{b} \; \bar{b}a. \tag{7.26}$$

None of these bad words appears on the right-hand side of (7.7), so the train track map is efficient.

7.5 Summary

In this chapter we gave an overview of train tracks and the Bestvina–Handel algorithm for finding the isotopy class of a map of D_n.

- Train tracks are oriented graphs, embedded in the surface D_n, with smoothness conditions (branched manifolds).
- A mapping class acts on a train track via its corresponding braid. The punctures are moved around and the train track is dragged along.
- The train track is chosen such that it remains invariant under this action, after being pushed back down to itself. We follow how edges are mapped to each other to create the train track map.
- If the train track was chosen properly, the train track map has no cancellations (i.e., no backtracking) in any of its iterates and is said to be efficient.
- The efficiency of the train track map allows us to compute the growth of edge lengths using simple linear algebra. We obtain a transition matrix that tells us how many times the image of an edge traces over an initial edge.
- In order for the isotopy class to be pseudo-Anosov, the transition matrix must be irreducible. The Perron–Frobenius theory then says that the matrix has a largest, unique real eigenvalue, which is the dilatation.
- The Bestvina–Handel algorithm systematically modifies the train track until it is found to be efficient, or we find a system of reducing curves.

For simple maps, such as the ones induced by taffy pullers [118], the invariant train track can usually be guessed at by examining the action on a few curves. The taffy puller maps often tend to be efficient, since cancellations are exactly what a taffy puller is designed to avoid.

The algorithm presented in this chapter constructs the entire train track and its associated map, which together carry a wealth of information about a pseudo-Anosov mapping, such as all its periodic orbits, and the transverse measure on each edge of the train track. However, if one is mostly interested in the dilatation of a pseudo-Anosov map, the algorithm is cumbersome enough that it is worth instead looking for a method that achieves less but obtains the dilatation more expediently. In the next chapter we will look at how we can iterate the Dynnikov coordinate representation of curves to find the dilatation very rapidly, but at the cost of not knowing the mapping class.

Chapter 8
Dynnikov Coordinates

"Do you understand how there could be any writing in a spider's web?" "Oh, no," said Dr. Dorian. "I don't understand it. But for that matter I don't understand how a spider learned to spin a web in the first place."

— *E. B. White,* Charlotte's Web

In Chap. 4, we gave a succinct description of any planar rod motion as a word in a braid group B_n. For $n = 3$, we showed in Chap. 3 that we could easily compute the topological entropy associated with the rod motion by multiplying matrices. However, this fails for $n > 3$ (except in special cases). There is a general technique due to Bestvina and Handel [12] that can be used for any n, but it is difficult to implement and computationally inefficient. In this chapter we introduce coordinates for describing non-oriented essential simple closed curves, known as *Dynnikov coordinates* [43]. These can represent arbitrarily convoluted closed curves by a small set of integers, though the integers themselves can get very large. Examining the growth of these integers under repeated applications of a braid allows us to compute the topological entropy for general braids. The coordinates also have other uses, such as a rapid solution to the word equality problem in B_n.

8.1 Coordinates for Multicurves

We call $\mathscr{S}(S)$ the space of equivalence classes under homotopy of non-oriented essential simple closed curves. We remind the reader that "essential" means that a curve is not contractible to a point, a puncture, or a boundary component. Thus, a curve that encircles all the punctures is not an element of $\mathscr{S}(D_n)$, since it is contractible to the outer boundary. An element of $\mathscr{S}(D_4)$ is depicted in Fig. 8.1 (top). Ideally, our goal would be to put coordinates on $\mathscr{S}(D_n)$, but in practice this turns out to be difficult. Surprisingly, a more "natural" space on which to define coordinates is $\mathscr{S}'(S)$, the space of *multicurves*. Multicurves consist of disjoint unions of essential simple closed curves, including ones where a component is "doubled-up" one or more times. (Multicurves are also called *integral laminations*.) A sample multicurve on D_4 is shown in Fig. 8.1 (bottom).

To show how we construct coordinates for $\mathscr{S}'(D_n)$, let us start with an example. Consider the non-oriented simple closed curve in Fig. 8.2, which is wrapped around $n = 5$ punctures (i.e., an element of $\mathscr{S}(D_5)$). As usual, by closed curve,

J.-L. Thiffeault, *Braids and Dynamics*, Frontiers in Applied Dynamical Systems: Reviews and Tutorials 9, https://doi.org/10.1007/978-3-031-04790-9_8

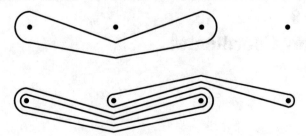

Fig. 8.1 Top: an element of the space $\mathscr{S}(D_4)$ of simple closed curve on the disk with 4 punctures. Bottom: an element of the space $\mathscr{S}'(D_4)$ of multicurves

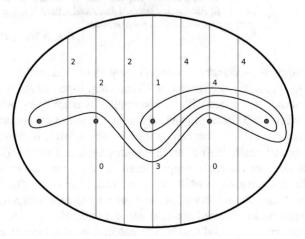

Fig. 8.2 A non-oriented simple closed curve wrapped around $n = 5$ punctures. In terms of the intersection numbers defined in Fig. 8.3, this curve has $v_1 = v_2 = 2$, $v_3 = v_4 = 4$, $\mu_1 = 2$, $\mu_3 = 1$, $\mu_5 = 4$, $\mu_2 = \mu_6 = 0$, $\mu_4 = 3$, and Dynnikov coordinate vector $u = (-1, 1, -2, 0, -1, 0)$ (see Eq. (8.3))

we really mean the equivalence class of the curve under homotopy. We have also drawn in Fig. 8.2 a set of vertical lines, where the two lines connected to each puncture are distinct. These lines are part of a topological *triangulation* of the punctured disk. (A true triangulation would also have lines connected to the first and last punctures, but these will not be needed here.) We count the minimum number of intersections between the closed curve and each vertical line, allowing homotopies of the curves if these can decrease the intersection number. It is a classical fact that we can reconstruct the isotopy class of a closed curve by counting how many times it intersects the fixed vertical lines in Fig. 8.2. (See for example the lectures by D. Thurston [124].) In Fig. 8.2 we can immediately see that the first puncture has a single curve segment folded around it, since the curve crosses the leftmost vertical line twice. We then proceed rightward and reconstruct as we go.

In Fig. 8.3 we define labels for the *intersection numbers* (also called crossing numbers in [117]). For n punctures, μ_{2i-3} (odd index) gives the number of intersections of a closed curve above the ith puncture and μ_{2i-2} (even index) below the same

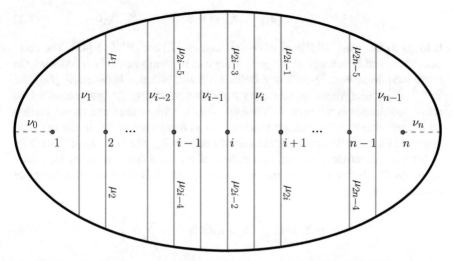

Fig. 8.3 Definition of the intersection numbers μ_i and ν_i. The μ_i for i odd count intersections above a puncture and below a puncture for i even. The ν_i count intersections between punctures

puncture. For $1 \leq i < n$, the number ν_i counts the intersections between punctures i and $i+1$. The intersection numbers ν_0 and ν_n are redundant, since for any curve we have

$$\nu_0 = \tfrac{1}{2}\nu_1, \qquad \nu_n = \tfrac{1}{2}\nu_{n-1}. \qquad (8.1)$$

It is convenient to define these redundant crossings for some calculations.

Not including ν_0 and ν_n, we thus have a total of $3n-5$ intersection numbers for a disk with n punctures. We make two observations. (i) An essential simple closed curve determines a set of intersection numbers, which can be used to uniquely reconstruct the curve, but in general an arbitrary collection of nonnegative integers does not determine such a curve. That is, intersection numbers obey some relations [46]. (ii) We can also reconstruct sets of disjoint closed curves from their total intersection numbers, that is, elements of $\mathscr{S}'(D_n)$.

There is thus a map from $\mathscr{S}'(D_n)$ to $\mathbb{Z}_{\geq 0}^{3n-5}$, but this map is not a bijection. This implies that we are carrying a lot of redundant information when using intersection numbers. To obtain a bijection, we define the differences

$$a_i = \tfrac{1}{2}(\mu_{2i} - \mu_{2i-1}), \qquad (8.2a)$$
$$b_i = \tfrac{1}{2}(\nu_i - \nu_{i+1}), \qquad (8.2b)$$

for $i = 1, \ldots, n-2$. The vector of length $(2n-4)$,[1]

---

[1] Some authors, such as Dynnikov [43] and Dehornoy [38], write the coordinate vector as $(a_1, b_1, \ldots, a_{n-2}, b_{n-2})$.

$$u = (a_1, \ldots, a_{n-2}, b_1, \ldots, b_{n-2}) \in \mathscr{D}_n(\mathbb{Z}) \approx \mathbb{Z}^{2n-4} \backslash \{0\}, \tag{8.3}$$

belongs to the space of *Dynnikov coordinates* $\mathscr{D}_n(\mathbb{Z})$ for $\mathscr{S}'(D_n)$ [43]. The components a_i and b_i are signed integers. They can be used to exactly reconstruct the curve [66]. Moreover, Dynnikov coordinates give a bijection between $\mathscr{S}'(D_n)$ and $\mathbb{Z}^{2n-4} \backslash \{0\}$ [66]. (We exclude 0 since we are not considering the "null curve.") This set of coordinates is minimal: it is not possible to achieve the same reconstruction with fewer numbers. Indeed, we need at least as many coordinates as the maximal degree of the dilatation for a pseudo-Anosov map on D_n, which is $2n - 4$ (Sect. 5.3).

To reconstruct the intersection numbers of a closed curve from its Dynnikov vector (8.3), we first solve for the end value v_1, which is obtained from the other coordinates as [66]

$$v_1 = 2 \max_{1 \le i \le n-2} \left(|a_i| + \max(b_i, 0) + \sum_{j=1}^{i-1} b_j \right). \tag{8.4}$$

We can now recover all the remaining v_i from (8.2b):

$$v_{i+1} = v_i - 2b_i, \quad i = 1, \ldots, n-2. \tag{8.5}$$

Now that we know the v_i, we can use them to obtain the μ_i. The intersection number μ_{2i} corresponds to a fixed line below puncture $i+1$, and v_i gives the number of intersections occurring between that puncture and the one on its left. We find

$$\mu_i = \begin{cases} (-1)^i a_{\lceil i/2 \rceil} + \frac{1}{2} v_{\lceil i/2 \rceil}, & b_{\lceil i/2 \rceil} \ge 0; \\ (-1)^i a_{\lceil i/2 \rceil} + \frac{1}{2} v_{1+\lceil i/2 \rceil}, & b_{\lceil i/2 \rceil} \le 0. \end{cases} \tag{8.6}$$

Later on we will need a measure of how convoluted a curve is. One simple measure is its minimum length: we pull the closed curve tight onto all the punctures, so that it consists of horizontal segments between punctures. We take the punctures to have zero size and be one unit of length apart. The total length $\ell(u)$ of the curve with Dynnikov coordinates u is then the sum of the number of segments between punctures, each given by v_i:

$$\ell(u) = \sum_{i=1}^{n-1} v_i. \tag{8.7}$$

This requires solving for the v_i first. Another measure of curve complexity, which is slightly simpler to compute, is the minimum number of intersections $L(u)$ of the closed curve with the horizontal line through the punctures [88]:

$$L(u) = |a_1| + |a_{n-2}| + \sum_{i=1}^{n-3} |a_{i+1} - a_i| + \sum_{i=0}^{n-1} |b_i|. \tag{8.8}$$

For example, the curve in Fig. 8.2 intersects the horizontal axis (the line through all the punctures) 12 times, so $L(u) = 12$. Coincidentally, the curve also has $\ell(u) = 2+2+4+4 = 12$.

8.2 Action of Braids on Dynnikov Coordinates (Update Rules)

In the previous section we discussed how the number of intersections with fixed reference lines (a triangulation) can be used to describe disjoint unions of essential simple closed curves. These (nonnegative) intersection numbers are highly redundant, since they must satisfy some relationships (such as triangle inequalities) if they are to correspond to essential simple closed curves. Passing to Dynnikov coordinates by taking differences of adjacent intersection numbers gives us a bijection between closed multicurves $\mathscr{S}'(D_n)$ and $\mathscr{D}_n(\mathbb{Z}) \approx \mathbb{Z}^{2n-4}\backslash\{0\}$. Some extra work is required to reconstruct the intersection numbers (and thus the curve) from Dynnikov coordinates, since ν_1 must be deduced from all the a_i and b_i via formula (8.4). This is the price to be paid for the concise description afforded by the Dynnikov coordinates.

The true advantage of the coordinates becomes more apparent when we examine the action of the braid group B_n on simple closed curves. Figure 8.4 illustrates the idea: we start from the same curve as in Fig. 8.2, and we apply a generator σ_1^{-1} to

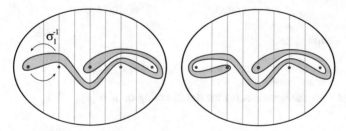

Fig. 8.4 Moving the punctures according to a braid generator changes some intersection numbers, in this case according to (8.20)

the first two punctures. The intersection numbers ν_1, μ_1, and μ_2 are modified, and the rest are unchanged by this operation. We could easily compute the intersection numbers before and after such an operation. However, since Dynnikov coordinates contain all the required information, it is also possible to work directly on the coordinates, as opposed to the intersection numbers. This leads to much more succinct expressions.

Given a curve encoded by u as in Eq. (8.3), each standard generator σ_i of the braid group B_n simply transforms these coordinates in a predetermined manner. This defines an *action* of the braid group on the set of Dynnikov coordinates. We call these transformations the *update rules* for a generator. Following Dehornoy [38], we write the action of a braid $\gamma \in B_n$ on a set of coordinates $u \in \mathscr{D}_n(\mathbb{Z})$ as a multiplicative right action:

$$\bar{u} = u \cdot \gamma. \tag{8.9}$$

To fully specify the action of B_n on $\mathscr{D}_n$, we need to give the update rules for the braid group generators σ_i and their inverses σ_i^{-1}. We can then compute the action of any braid by using the group homomorphism property, e.g., $u \cdot (\sigma_1 \sigma_2) = (u \cdot \sigma_1) \cdot \sigma_2$. Notice that we use a right action, since we read braids left to right.

The formulas for the update rules are obtained by a straightforward but computationally lengthy calculation. We relegate the full derivation to Appendix A, which is based on notes by Spencer A. Smith, who bases his derivation on Whitehead moves for the triangulation defined in Fig. 8.3.

8.2.1 Update Rules for σ_i

We give the update rules for the braid group generators

$$\bar{u} = u \cdot \sigma_i, \qquad i = 1, \dots, n-1. \tag{8.10}$$

To express the update rules succinctly,[2] first define for a quantity f the operators

$$f^+ := \max(f, 0), \qquad f^- := \min(f, 0). \tag{8.11}$$

After we define

$$c_{i-1} = a_{i-1} - a_i - b_i^+ + b_{i-1}^-, \tag{8.12}$$

we can express the update rules for the σ_i acting on $u = (a_1, \dots, a_{n-2}, b_1, \dots, b_{n-2})$ as

$$\bar{a}_{i-1} = a_{i-1} - b_{i-1}^+ - \left(b_i^+ + c_{i-1}\right)^+, \tag{8.13a}$$

$$\bar{b}_{i-1} = b_i + c_{i-1}^-, \tag{8.13b}$$

$$\bar{a}_i = a_i - b_i^- - \left(b_{i-1}^- - c_{i-1}\right)^-, \tag{8.13c}$$

$$\bar{b}_i = b_{i-1} - c_{i-1}^-, \tag{8.13d}$$

for $i = 2, \dots, n-2$. For this and the following update rules, all the other unlisted components of u are unchanged under the action of σ_i or σ_i^{-1}.

The update rules (8.13) can also be used for the leftmost ($i = 1$) and rightmost ($i = n-1$) generators, as long as we naturally extend the Dynnikov coordinates by

---

[2] We are using the numbering scheme of [66], but the notation of [88]. Also, we define generators σ_i as clockwise interchanges rather than counterclockwise.

$$a_0 = a_{n-1} = 0, \quad b_0 = -\tfrac{1}{2}v_1, \quad b_{n-1} = \tfrac{1}{2}v_{n-1}. \tag{8.14}$$

In theory, the update rules (8.13) for $i = 1$ or $i = n - 1$ involve the quantities b_0 and b_{n-1}, which are not themselves Dynnikov coordinates and must be obtained from (8.4) and (8.5). However, they magically disappear after exploiting the $^+$ and $^-$ operators in the following manner. Geometrically, $\mu_1 + \mu_2 = v_1 = -2b_0$, and $-\mu_1 + \mu_2 = 2a_1$ from the definition (8.2a). Summing these gives $a_1 - b_0 = \mu_2 \geq 0$, so that, for instance, $c_0^- = (a_0 - a_1 - b_1^+ + b_0^-)^- = (-a_1 + b_0 - b_1^+)^- = -a_1 + b_0 - b_1^+$, since the argument of $^-$ is negative. From (8.13d) with $i = 1$, we thus get $\bar{b}_1 = a_1 + b_1^+$, with no dependence on the extended variable b_0. After some simplifications, we thus obtain from (8.13c)–(8.13d) with $i = 1$ the update rules for σ_1:

$$\bar{a}_1 = -b_1 + \left(a_1 + b_1^+\right)^+, \tag{8.15a}$$
$$\bar{b}_1 = a_1 + b_1^+. \tag{8.15b}$$

An analogous simplified update rule for σ_{n-1} is also obtained from (8.13a)–(8.13b) with $i = n - 1$:

$$\bar{a}_{n-2} = -b_{n-2} + \left(a_{n-2} + b_{n-2}^-\right)^-, \tag{8.16a}$$
$$\bar{b}_{n-2} = a_{n-2} + b_{n-2}^-. \tag{8.16b}$$

8.2.2 Update Rules for σ_i^{-1}

We need to give separate update rules for the generators σ_i^{-1},

$$\bar{u} = u \cdot \sigma_i^{-1}, \qquad i = 1, \ldots, n - 1, \tag{8.17}$$

which of course are the inverses of the previous update rules. With the definition

$$d_{i-1} = a_{i-1} - a_i + b_i^+ - b_{i-1}^-, \tag{8.18}$$

the update rules for the σ_i^{-1} are

$$\bar{a}_{i-1} = a_{i-1} + b_{i-1}^+ + \left(b_i^+ - d_{i-1}\right)^+, \tag{8.19a}$$
$$\bar{b}_{i-1} = b_i - d_{i-1}^+, \tag{8.19b}$$
$$\bar{a}_i = a_i + b_i^- + \left(b_{i-1}^- + d_{i-1}\right)^-, \tag{8.19c}$$
$$\bar{b}_i = b_{i-1} + d_{i-1}^+, \tag{8.19d}$$

for $i = 2, \ldots, n - 2$. We also have

$$\bar{a}_1 = b_1 - \left(b_1^+ - a_1\right)^+, \tag{8.20a}$$
$$\bar{b}_1 = b_1^+ - a_1, \tag{8.20b}$$

for σ_1^{-1}, and

$$\bar{a}_{n-2} = b_{n-2} - \left(b_{n-2}^- - a_{n-2}\right)^-, \tag{8.21a}$$
$$\bar{b}_{n-2} = b_{n-2}^- - a_{n-2}, \tag{8.21b}$$

for σ_{n-1}^{-1}.

Update rules of this form are known as *piecewise-linear*: once the mins and maxes are resolved, what is left is simply a linear operation. In the next section we give a different form of the update rules and discuss how the update rules are related to each other.

8.3 Max-Plus Algebra

A convenient way of representing the action of braids on Dynnikov coordinates (the update rules from the previous section) is through a *max-plus algebra* (or max-plus semiring) [31, 71, 85]. This is an algebra containing two operations, $\oplus$ and $\otimes$, defined as

$$x \oplus y = \max(x, y), \qquad x \otimes y = x + y, \tag{8.22}$$

where $x, y \in \mathbb{R} \cup \{-\infty\}$ are two arbitrary numbers. Strangely, addition $+$ becomes a kind of *multiplication* $\otimes$ in the max-plus algebra. Both operations $\oplus$ and $\otimes$ are associative and commutative. The element $\mathbb{O} = -\infty$ plays the role of zero:

$$\mathbb{O} \oplus x = x, \qquad \mathbb{O} \otimes x = \mathbb{O}, \qquad \text{with } \mathbb{O} := -\infty. \tag{8.23}$$

The element $\mathbb{I} = 0$ is the multiplicative unit:

$$\mathbb{I} \otimes x = x, \qquad \text{with } \mathbb{I} := 0. \tag{8.24}$$

A striking fact is that these two operations are *distributive*:

$$(x \oplus y) \otimes z = (x \otimes z) \oplus (y \otimes z). \tag{8.25}$$

In terms of the original operations, the distributivity property reads

$$\max(x, y) + z = \max(x + z, y + z), \tag{8.26}$$

which is obvious, since adding the same factor to both operands of max does not change the outcome.

The max-plus algebra is different from regular operations over the reals in several ways. Addition is idempotent, meaning that adding an element to itself returns the same element:

$$x \oplus x = x. \tag{8.27}$$

There is a multiplicative inverse, since

$$x \otimes (-x) = \mathbb{I}, \qquad x^{-1} = -x. \tag{8.28}$$

However, there is no additive inverse.

We now follow a similar notation to Hall and Yurttaş [66] and use double brackets $[\![\cdot]\!]$ to indicate when addition and multiplication should be interpreted as taking place in the max-plus algebra. We thus write

$$x \oplus y = [\![x+y]\!], \quad x \otimes y = [\![xy]\!], \quad [\![1]\!] = 0, \quad [\![x/y]\!] = x-y. \tag{8.29}$$

Now we can rewrite the update rules (8.15) as

$$\bar{a}_1 = \left[\!\!\left[\frac{1+a_1(1+b_1)}{b_1}\right]\!\!\right], \qquad \bar{b}_1 = [\![a_1(1+b_1)]\!], \tag{8.30}$$

the update rules (8.13) as

$$\bar{a}_{i-1} = \left[\!\!\left[\frac{a_{i-1}a_i}{a_{i-1}b_{i-1}+a_i(1+b_{i-1})}\right]\!\!\right], \quad \bar{b}_{i-1} = \left[\!\!\left[\frac{a_{i-1}b_{i-1}b_i}{a_{i-1}b_{i-1}+a_i(1+b_{i-1})}\right]\!\!\right] \tag{8.31a}$$

$$\bar{a}_i = \left[\!\!\left[\frac{a_{i-1}+a_i(1+b_i)}{b_i}\right]\!\!\right], \qquad \bar{b}_i = \left[\!\!\left[\frac{a_{i-1}b_{i-1}+a_i(1+b_{i-1})(1+b_i)}{a_{i-1}}\right]\!\!\right] \tag{8.31b}$$

for $i = 2,\ldots,n-2$, and the update rules (8.16) as

$$\bar{a}_{n-2} = \left[\!\!\left[\frac{a_{n-2}}{a_{n-2}b_{n-2}+1+b_{n-2}}\right]\!\!\right], \qquad \bar{b}_{n-2} = \left[\!\!\left[\frac{a_{n-2}b_{n-2}}{1+b_{n-2}}\right]\!\!\right]. \tag{8.32}$$

For completeness, we also give the update rules for the inverse generators. The update rules (8.20) are rewritten as

$$\bar{a}_1 = \left[\!\!\left[\frac{a_1 b_1}{a_1+1+b_1}\right]\!\!\right], \qquad \bar{b}_1 = \left[\!\!\left[\frac{1+b_1}{a_1}\right]\!\!\right], \tag{8.33}$$

the update rules (8.19) as

$$\bar{a}_{i-1} = [\![a_{i-1}(1+b_{i-1})+a_ib_{i-1}]\!], \quad \bar{b}_{i-1} = \left[\!\!\left[\frac{a_ib_{i-1}b_i}{a_{i-1}(1+b_{i-1})(1+b_i)+a_ib_{i-1}}\right]\!\!\right]$$

$$(8.34\text{a})$$

$$\bar{a}_i = \left[\!\!\left[\frac{a_{i-1}a_ib_i}{a_{i-1}(1+b_i)+a_i}\right]\!\!\right], \qquad \bar{b}_i = \left[\!\!\left[\frac{a_{i-1}(1+b_{i-1})(1+b_i)+a_ib_{i-1}}{a_i}\right]\!\!\right]$$

$$(8.34\text{b})$$

for $i = 2,\ldots,n-2$, and the update rules (8.21) as

$$\bar{a}_{n-2} = [\![a_{n-2}(1+b_{n-2})+b_{n-2}]\!], \qquad \bar{b}_{n-2} = \left[\!\!\left[\frac{b_{n-2}}{a_{n-2}(1+b_{n-2})}\right]\!\!\right]. \qquad (8.35)$$

We remind the reader that these update rules are identical to [66], with σ_i and σ_i^{-1} interchanged, since we define positive generators clockwise.

The relationship between the update rule for σ_i and that for σ_i^{-1} can be understood as follows. The operation

$$(a_1,\ldots,a_{n-2},b_1,\ldots,b_{n-2}) \quad \longmapsto \quad (-a_1,\ldots,-a_{n-2},b_1,\ldots,b_{n-2}) \qquad (8.36)$$

corresponds to a vertical reflection of a closed curve (i.e., with respect to the horizontal axis). This is an involution, since doing it twice gives the identity. In the max-plus notation, the involution replaces $[\![a_i]\!]$ by $[\![1/a_i]\!]$. To obtain the update rule for σ_i^{-1} from that for σ_i, we apply the involution to the coordinates, then apply the update rule for σ_i, and finally apply the involution again. This corresponds to conjugation by the involution.

Another interesting involution is horizontal reflection, which has

$$(a_1,\ldots,a_{n-2},b_1,\ldots,b_{n-2}) \quad \longmapsto \quad (a_1,\ldots,a_{n-2},-b_1,\ldots,-b_{n-2}). \qquad (8.37)$$

Conjugating by this involution relates the update rule for σ_i to that of σ_{n-i-1}^{-1}. For example, consider the update rule (8.33) for σ_1^{-1}; applying the sequence involution-update rule-involution for b_1 gives

$$\left[\!\!\left[\left(\frac{1+1/b_1}{a_1}\right)^{-1}\right]\!\!\right] = \left[\!\!\left[\left(\frac{b_1+1}{a_1b_1}\right)^{-1}\right]\!\!\right] = \left[\!\!\left[\frac{a_1b_1}{1+b_1}\right]\!\!\right], \qquad (8.38)$$

which is exactly the update rule of b_{n-2} for σ_{n-2} in Eq. (8.32), with $n-2$ replaced by 1. This example shows how much easier it is to manipulate max-plus expressions, which use a familiar algebra, compared to directly working with expressions involving min and max. This can sometimes allow an explicit computation of the action of a braid on Dynnikov coordinates, as done in Hall and Yurttaş [66].

8.4 Mapping Classes and Dynnikov Coordinates

Recall that in Chap. 5, we discussed the Thurston–Nielsen classification of diffeomorphisms (Theorem 5.1). For diffeomorphisms of the n-punctured disk D_n, we can express the three different cases of the classification by their action on Dynnikov coordinates.

Let $\phi \in \text{Diff}^+(D_n)$ be a representative of an element of the mapping class group $\text{MCG}(D_n)$. As we saw in Chap. 4, an element of the mapping class group can be written as a braid $\text{br}(\phi) \in B_n$. We can then consider the action of the braid $\text{br}(\phi)$ on elements of $\mathscr{S}_n$. (We are really considering elements of $B_n/\langle\Delta_n^2\rangle$, since the full twist Δ_n^2 acts trivially on $\mathscr{S}_n$; see (4.8).)

8.4.1 Finite-Order Case

If ϕ belongs to a finite-order mapping class, then there exists an integer $k > 0$ such that

$$u \cdot [\text{br}(\phi)]^k = u, \qquad \forall u \in \mathscr{D}_n(\mathbb{Z}). \tag{8.39}$$

That is, $[\text{br}(\phi)]^k$ acts trivially on $\mathscr{D}_n(\mathbb{Z})$.

Example: One great advantage of the max-plus formulation of Sect. 8.3 is that it is well-suited to symbolic manipulation with software such as *Mathematica*. Let $n = 3$ and take $\text{br}(\phi) = \sigma_1\sigma_2$. We can code the update rules (8.30) and (8.32) (with $n = 3$) as two functions sig1 and sig2:

```
In[1]:= sig1[{a_,b_}] := {(1+a(1+b))/b, a(1+b)}

In[2]:= sig2[{a_,b_}] := {a/(a b+1+b), a b/(1+b)}

In[3]:= f[u_] := Simplify[sig2[sig1[u]]]

In[4]:= NestList[f, {a, b}, 3]

Out[4]= {{a, b},
         {1/(a + b + a b), a (1 + 1/b)},
         {b/(1 + a + a b), 1/(a + a b)},
         {a, b}}
```

In the last step NestList returns a list containing $\{f^0(u), f^1(u), f^2(u), f^3(u)\}$, where f is the action of $\sigma_1\sigma_2$ on the Dynnikov coordinates $u = \{a_1, b_1\} = \{\text{a}, \text{b}\}$. The Dynnikov coordinates thus always return to their initial value after $k = 3$ iterates. This shows that the braid $\sigma_1\sigma_2$ is finite-order. Keep in mind, though, that the validity of this particular implementation is limited, since the idempotent property (8.27) is not accounted for.

8.4.2 Reducible Case

If ϕ belongs to a reducible mapping class, then

$$u_0 \cdot \mathrm{br}(\phi) = u_0, \qquad \text{for some } u_0 \in \mathscr{D}_n(\mathbb{Z}), \tag{8.40}$$

but such that $\mathrm{br}(\phi)$ does not act trivially on $\mathscr{D}_n(\mathbb{Z})$. Here u_0 is not unique: a multiple $c u_0$ for some $c \in \mathbb{Z}$, $c \neq 0$, will also satisfy (8.40). If u_0 describes a disjoint union of curves, then each connected component may satisfy (8.40), unless the components are permuted by $\mathrm{br}(\phi)$. The vector u_0 describes the system of curves Γ in the statement of Theorem 5.1, for the reducible case.

Example: Take the braid $\sigma_1 \sigma_3$ in B_4. Since this braid pairwise-interchanges punctures 1 and 2 and punctures 3 and 4, we expect reducing curves around each pair. Indeed, we have

$$(0,0,1,-1) \cdot (\sigma_1 \sigma_3) = (0,0,1,-1), \tag{8.41}$$

where $(0,0,1,-1)$ is the Dynnikov coordinate vector for the multicurve in Fig. 8.5.

In this case, the two components of the multicurve are not permuted. We thus expect each individual curve to be invariant:

$$(0,0,1,0) \cdot (\sigma_1 \sigma_3) = (0,0,1,0), \qquad (0,0,0,-1) \cdot (\sigma_1 \sigma_3) = (0,0,0,-1). \tag{8.42}$$

The half-twist braid $\Delta_4 = \sigma_3 \sigma_2 \sigma_1 \sigma_3 \sigma_2 \sigma_3$ (Eq. (4.6)) also has $(0,0,1,-1)$ as a system of reducing curves, but in this case the two components are permuted:

$$(0,0,1,0) \cdot \Delta_4 = (0,0,0,-1), \qquad (0,0,0,-1) \cdot \Delta_4 = (0,0,1,0). \tag{8.43}$$

Example: The reducing curves in Fig. 7.12 are permuted by the reducible braid γ defined in (7.12). We can see this by considering the components individually:

$$u_1 = (0,0,0,0,1,0,0,0), \quad u_2 = (0,0,-1,0,0,-1,0,1),$$
$$u_3 = (0,0,0,1,0,0,-1,0) \tag{8.44}$$

and then observing that $u_1 \cdot \gamma = u_2$, $u_2 \cdot \gamma = u_3$, $u_3 \cdot \gamma = u_1$. Note that in this case, since the curves are disjoint, the Dynnikov coordinates for the system of reducing curves is $u = u_1 + u_2 + u_3$, with $u \cdot \gamma = u$.

8.4.3 Pseudo-Anosov Case

The pseudo-Anosov case of Theorem 5.1 is the most interesting, but it also requires special interpretation when dealing with Dynnikov coordinates. Let us illustrate the issue with an example, the braid $\gamma = \sigma_1 \sigma_2 \sigma_3^{-1}$, whose action on a material line in a fluid is depicted in Fig. 5.7.

Figure 8.6 shows the action of γ on the curve $u = (0,0,1,-1)$ of Fig. 8.5. The curve rapidly becomes extremely complicated, and we run out of resolution. This can be seen in the increase of the numeric values themselves:

Fig. 8.5 System of reducing curves for the braids $\sigma_1 \sigma_3$ and Δ_4

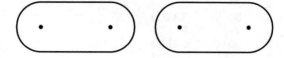

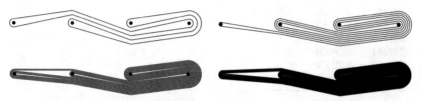

Fig. 8.6 Four iterates of the action of $\sigma_1 \sigma_2 \sigma_3^{-1}$ on the curve $(0,0,1,-1)$ of Fig. 8.5

$$
\begin{aligned}
u \cdot \gamma &= (-1,2,-1,-1) \\
u \cdot \gamma^2 &= (1,4,-3,-3) \\
u \cdot \gamma^3 &= (3,8,-3,-7) \\
u \cdot \gamma^4 &= (3,20,-9,-15) \\
u \cdot \gamma^5 &= (9,46,-23,-35).
\end{aligned}
\tag{8.45}
$$

Let us measure the growth by taking the length $\ell(u)$ defined in Eq. (8.7). Table 8.1 shows how the length $\ell(u \cdot \gamma^k)$ increases with k. As expected, it grows rapidly. However, in the last column, we take the ratio of two successive iterates of the length: this quantity appears to converge. Indeed, for a pseudo-Anosov map, we have

$$
\lambda = \exp(h(\gamma)) = \lim_{k \to \infty} \frac{\ell(u \cdot \gamma^k)}{\ell(u \cdot \gamma^{k-1})} = \lim_{k \to \infty} \left(\ell(u \cdot \gamma^k) \right)^{1/k},
\tag{8.46}
$$

where λ is the dilatation of a pseudo-Anosov map (see Theorem 5.1) associated with the braid γ. The topological entropy is $h(\gamma) = \log \lambda$.

This iterative method is a fast way of computing the topological entropy (or dilatation). The answer in this case is $\lambda = 2.2966305266\ldots$, so we are getting about five digits of precision in 15 iterations. This is in agreement with the characteristic polynomial (7.10) for the same braid: its second root is on the unit circle, so we should be getting about $\log_{10}(2.2966) \approx 0.36$ digits per iterate, which translates

Table 8.1 Growth of $\ell(u)$ for the braid $\gamma = \sigma_1 \sigma_2 \sigma_3^{-1}$

k	$\ell(u \cdot \gamma^k)$	$\ell(u \cdot \gamma^k)/\ell(u \cdot \gamma^{k-1})$
0	4	–
1	12	3
2	24	2
3	56	2.33333
4	132	2.35714
5	300	2.27273
6	688	2.29333
7	1584	2.30233
8	3636	2.29545
9	8348	2.29593
10	19176	2.29708
11	44040	2.29662
12	101140	2.29655
13	232284	2.29666
14	533472	2.29664
15	1225184	2.29662

into about 14 iterations for 5 digits. Note that we could have used the intersection number $L(u)$ (Eq. (8.8)) instead of $\ell(u)$ and obtain the same results [46, 88, 117].

The convergence occurs in the same manner as Lyapunov exponents in the general ergodic theory of dynamical systems [95]. The rate of convergence will depend on the difference between the two largest Lyapunov exponents. Moussafir [88] has a conjecture on the rate of convergence that was recently disproved by Bell and Schleimer [11]. They showed that there exist braids for which the two leading Lyapunov exponents are arbitrarily close, so that no guarantees can be made on the rate of convergence of the limit in (8.46). Nevertheless, in practice approximating (8.46) is a very rapid method of computing the topological entropy or dilatation for a typical braid. More details of the practical aspects of the computation are given in [29, 117, 119], as well as in [49] for a related triangulation on the torus.

To deal with pseudo-Anosov braids, it is natural to extend the Dynnikov coordinates to the real numbers, which we write as $\mathscr{D}_n(\mathbb{R})$. In that case, if $\mathrm{br}(\phi)$ is the braid corresponding to a pseudo-Anosov map, we have

$$u_\mathrm{u} \cdot \mathrm{br}(\phi) = \lambda u_\mathrm{u}, \qquad u_\mathrm{s} \cdot \mathrm{br}(\phi) = \lambda^{-1} u_\mathrm{s}, \tag{8.47}$$

where u_u corresponds to the unstable foliation $\mathscr{F}_\mathrm{u}$, and u_s corresponds to the stable foliation $\mathscr{F}_\mathrm{s}$. The real-valued Dynnikov vector then contains information about the foliation and its measure. If we reconstruct the intersection numbers μ_i and ν_i, they will correspond to the positive weights associated with intersections with the Dynnikov triangulation. Hall and Yurttaş [66] have exploited the Dynnikov coordinate formulation to explicitly find large families of pseudo-Anosov braids.

8.5 The Word Problem

A classical problem in the algebraic theory of braids is how to determine equality of two braids, given as products of generators σ_i and their inverses (a *word*). This is called the "word problem." This is harder than it sounds, since multiple applications of the relations (4.5) may be required before the two braids are manifestly equal. See the review by Birman and Brendle [19] for the various methods that were proposed over the years.

As originally pointed out by Dynnikov [43] (see also Dehornoy [38] for a review), the coordinates can be used to solve the word problem very rapidly. We consider the curve

$$u = (0, 0, \ldots, 0, -1, -1, \ldots, -1) \in \mathscr{D}_{n+1}(\mathbb{Z}). \qquad (8.48)$$

That is, all the a_i are zero and all the b_i are set to -1. We take this curve to be in $\mathscr{D}_{n+1}(\mathbb{Z})$, which means we add an extra puncture, shown as a hollow dot in Fig. 8.7.

Fig. 8.7 The closed multicurve $(0,0,0,0,-1,-1,-1,-1)$ in $\mathscr{D}_{5+1}(\mathbb{Z})$

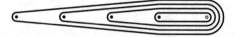

Now take an arbitrary braid $\gamma \in B_n$. This braid acts naturally on B_{n+1}, basically by leaving the rightmost puncture alone. But the crucial fact is that γ must modify the multicurve (8.48), unless γ is the trivial braid. We added the extra puncture so that full twists Δ_n^2 (see Eq. (4.6)) would not act trivially. Hence, to determine if two braids γ and γ' are equal, simply act on u with $\gamma^{-1}\gamma'$, and see if $u \cdot (\gamma^{-1}\gamma') = u$. Equivalently, one can check $u \cdot \gamma = u \cdot \gamma'$. This is a geometrically elegant and rapid solution to the word problem. Unfortunately, this does not help to solve the much harder problem of determining if two braids are conjugate to each other.

8.6 Summary

Dynnikov coordinates are a powerful tool for studying curves on punctured disks and how these curves are affected by dynamics.

- Because non-intersecting closed curves on a surface are very tightly constrained, they can be reconstructed from their minimum intersection numbers with fixed curves (i.e., a triangulation).
- By taking differences of these intersection numbers, we get a bijection between equivalence classes of multicurves and $\mathbb{Z}^{2n-4}$ — the Dynnikov coordinates.
- The braid group acts on Dynnikov coordinates by a piecewise-linear action. This is very fast and easy to implement numerically. The action gives a straightforward

solution to the word problem in the braid group (i.e., deciding equality of two braid words).

- The action on Dynnikov coordinates is used by software such as braidlab [119] to numerically compute entropy for very large braids.

The Dynnikov coordinates have been very useful for analyzing braids of large datasets. Applications range from fluid dynamics [3, 29, 32, 48, 56, 115], to crowd dynamics [2, 84], granular media [100], and flocking of birds [34]. There are also other promising approaches that may generalize better to higher dimensions [103].

Chapter 9
The Braidlab Library

Braidlab is a software library for manipulating braids and loops, written in Matlab by the author and Marko Budišić. Unlike many software packages for exploring braids, braidlab was written to be fast and to scale to very large braids, involving millions of strings. We will see in Chap. 10 applications involving large braids, and in the present chapter we give a brief tour of braidlab.

9.1 Setup and Getting Help

The first step in using braidlab is to have Matlab installed on your system. The second step is to download the braidlab package from https://github.com/jeanluct/braidlab/releases/ and extract the archive file to a directory on your computer, which we shall symbolically refer to as BRAIDLAB_DIRECTORY. To make sure that we can access the braidlab commands, its directory must be known to Matlab. This can be done within Matlab by executing

```
>> addpath BRAIDLAB_DIRECTORY
>> import braidlab.*
```

The `import` command allows us to subsequently type braidlab commands without an explicit `braidlab.` prefix. (This prefix is called a *namespace*.)

In object-oriented languages such as Matlab, mathematical concepts like braids and loops are conveniently represented by software objects called *classes*. Classes have their own *methods*, which are functions that manipulate a particular class. In particular, there is a distinguished method called a *constructor* that is used to create instances of that class. One of the basic building blocks in braidlab is the `braid` class, which is used to represent a simple braid on a disk. The command `help braid` describes the class `braid`:

© The Author(s), under exclusive license to Springer Nature Switzerland AG 2022
J.-L. Thiffeault, *Braids and Dynamics*, Frontiers in Applied Dynamical Systems:
Reviews and Tutorials 9, https://doi.org/10.1007/978-3-031-04790-9_9

```
>> help braid

 braid   Class for representing braids.
    A braid object holds a braid represented in terms of Artin
    generators.

    The class braid has the following data members:

     'word'       vector of signed integers (int32) giving the Artin
                  generators
     'n'          number of strings in the braid

    METHODS('braid') shows a list of methods.

    See also braid/braid (constructor), cfbraid.

    Documentation for braidlab.braid
```

To get more information on the `braid` constructor, invoke

```
>> help braid.braid
 braid   Construct a braid object.
    B = braid(W) creates a braid object B from a vector of
    generators W.

    B = braid(W,N) specifies the number of strings N of the
    braid group, which is otherwise guessed from the maximal
    elements of W.

    The braid group generators are represented as a list of
    integers I satisfying -N < I < N.  The usual group operations
    (multiplication, inverse, powers) can be performed on braids.
    ...
```

which refers to the method `braid` (the constructor) within the class `braid`. Use

```
>>   methods(braid)
```

to list all the methods in the class:

```
Methods for class braidlab.braid:

alexpoly   compact   entropy     ispure     lk          ne
tensor     braid     complexity  eq         istrivial   loopcoords
perm       train     burau       conjtest   gencount    length
mpower     plot      writhe      char       cycle       inv
lexeq      mtimes    subbraid
```

As mentioned above, methods are functions that are tied to a particular class, in this case `braid`, and can only act on objects of that class type. For example, given a braid b, we can invoke the `inv` (inverse) method with either `inv(b)` or `b.inv`. The special methods `mtimes` and `mpower` are automatically called when the multiplication * and power ^ operators are used on `braids`. The special method `eq` is called when the equality test == is performed on `braids`.

9.2 Braids

The braid class represents a simple algebraic Artin braid on a disk with n punctures. We summarize the basic creation and manipulation of braids.

9.2.1 Basic Operations

The braid $\sigma_1 \sigma_2^{-1}$ is constructed with

```
>> a = braid([1 -2])    % defaults to 3 strings

a = < 1 -2 >
```

The angle brackets serve to distinguish the output from a regular Matlab array. By default, braidlab sets the number of strings based on the generators used. The same braid on 4 strings is constructed with

```
>> a4 = braid([1 -2],4)    % force 4 strings

a4 = < 1 -2 >
```

Two braids can be multiplied (joined together):

```
>> a = braid([1 -2]); b = braid([1 2]);
>> a*b, b*a

ans = < 1 -2  1  2 >

ans = < 1  2  1 -2 >
```

Powers can also be taken, including the inverse:

```
>> a^5, inv(a), a*a^-1

ans = < 1 -2  1 -2  1 -2  1 -2  1 -2 >

ans = < 2 -1 >

ans = < 1 -2  2 -1 >
```

This last expression is the identity braid but is not simplified. The method compact attempts to simplify the braid:

```
>> compact(a*a^-1)

ans = < e >
```

The method compact is based on the heuristic algorithm of Bangert et al. [10], since finding the braid of minimum length in the standard generators is in general difficult [97]. Hence, there is no guarantee that in general compact will find the identity braid, even though it does so here. To really test if a braid is the identity (trivial braid), use the method `istrivial`:

```
>> istrivial(a*a^-1)

ans = 1
```

The number of strings for an existing braid a can be found with

```
>> a.n

ans = 3
```

For convenience, braidlab allows the multiplication of braids with a different number of strings. If two such braids are multiplied, then the resulting braid has the maximum number of strings of the two:

```
>> a = braid([1 -2]); b = braid([1 2 -3]);
>> c = a*b  % a 3-string braid times a 4-string braid

c = < 1 -2  1  2 -3 >

>> c.n

ans = 4
```

There are other ways to construct a `braid`, such as using random generators, here a braid with 5 strings and 10 random generators:

```
>> braid('Random',5,10)

ans = < 1  4 -4  2  4 -1 -2  4  4  4 >
```

The constructor can also build some standard braids:

```
>> braid('HalfTwist',5)

ans = < 4  3  2  1  4  3  2  4  3  4 >

>> braid('8_21')   % braid for 8-crossing knot #21

ans = < 2  2  2  1  2  2 -1 -1  2 -1 >
```

In Chap. 10 we will show how to construct a braid from a trajectory dataset.

The `braid` class handles equality of braids:

```
>> a = braid([1 -2]); b = braid([1 -2 2 1 2 -1 -2 -1]);
>> a == b

ans = 1
```

Using the braid relations of Chap. 4, we can easily see that these are the same braid, even though they appear different from their generator sequence. Equality is determined efficiently by acting with the braids on Dynnikov coordinates [43], as described in Chap. 8; see Sect. 9.3.2 below for more details. If for some reason lexicographic (generator-per-generator) equality of braids is needed, use the method `lexeq(b1,b2)`.

We can extract a subbraid by choosing specific strings: for example, if we take the 4-string braid $\sigma_1\sigma_2\sigma_3^{-1}$ and discard the third string, we obtain $\sigma_1\sigma_2^{-1}$, as depicted in Fig. 9.1:

```
>> a = braid([1 2 -3]);
>> subbraid(a,[1 2 4])    % subbraid using strings 1,2,4

ans = < 1 -2 >
```

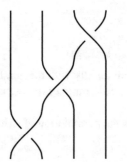

Fig. 9.1 Removing the third string from the braid $\sigma_1\sigma_2\sigma_3^{-1}$ (left) yields the braid $\sigma_1\sigma_2^{-1}$ (right)

The *tensor product* of two braids is the larger braid obtained by laying two braids side by side [72]:

```
>> a = braid([1 2 -3]); b = braid([1 -2]);
>> tensor(a,b)

ans = < 1  2  -3  5  -6 >
```

Here, the tensor product of a 4-braid and a 3-braid has 7 strings. The generators $\sigma_1\sigma_2^{-1}$ of b become $\sigma_5\sigma_6^{-1}$ after re-indexing, so they appear to the right of a.

9.2.2 Representation and Invariants

There are a few remaining methods in the braid class, which we describe briefly. The reduced Burau matrix representation [18, 30] of a braid (Sect. 4.5) is obtained with the method `burau`:

```
>> burau(braid([1 -2]),-1)

ans = 1    -1
     -1     2
```

where the last argument (-1) is the value of the parameter t in the Laurent polynomials that appear in the entries of the Burau matrices. We can use Matlab's symbolic toolbox to specify the Burau parameter as a variable:

```
>> B = burau(braid([1 -2]),sym('t'))

B = [ -t,        t]
    [ -1, 1 - 1/t]
```

Another well-known homological representation of braid groups is the Lawrence–Krammer representation [14, 78], which we discussed at the end of Chap. 4. It is given in terms of two parameters, usually denoted t and q:

```
>> K = lk(braid([1 -2]),sym('t'),sym('q'))

K = [ - (q-1)^2/q - q*t*(q-1),  - (q-1)/q,  (q^2-q+1)/q^2]
    [             -q^2*t,             0,               0]
    [      - (q-1)/(q*t),     -1/(q*t),   (q-1)/(q^2*t)]
```

Because its size grows more rapidly with the number of strings (matrices of dimension $\frac{1}{2}n(n-1)$), the Lawrence–Krammer representation is very slow to compute for large braids.

The method perm gives the permutation of strings corresponding to a braid:

```
>> perm(braid([1 2 -3]))

ans = 2   3   4   1
```

If the strings are unpermuted, then the braid is *pure*, which can also be tested with the method ispure.

Finally, the *writhe* of a braid is the sum of the powers of its generators. The writhe of $\sigma_1^{+1}\sigma_2^{+1}\sigma_3^{-1}$ is $+1+1-1 = 1$:

```
>> writhe(braid([1 2 -3]))

ans = 1
```

The writhe is a braid invariant.

9.3 Loops

Dynnikov coordinates for equivalence classes of loops, also known as loop coordinates, were described in Chap. 8. We considered equivalence classes of such loops under homotopies relative to the punctures. In particular, the loops are essential, meaning that they are not null-homotopic or homotopic to the boundary or a puncture. The intersection numbers shown in Fig. 8.2 (page 80) count the minimum number of intersections of an equivalence class of loops with the fixed vertical lines shown. For n punctures, we define the intersection numbers μ_i and ν_i in Fig. 8.3.

Let us create the loop in Fig. 8.2 as a loop object:

```
>> u = loop([-1 1 -2 0 -1 0])

u = (( -1 1 -2 0 -1 0 ))
```

The double parentheses indicate a loop object. The coordinates are ordered as for the vector u in Eq. (8.3). We can convert from loop coordinates to intersection numbers with

```
>> intersec(u)

ans = 2 0 1 3 4 0 2 2 4 4    % [mu1 ... mu6 nu1 ... nu4]
```

which returns $\mu_1 \ldots \mu_{2n-4}$ followed by $\nu_1 \ldots \mu_{n-1}$, as defined in Fig. 8.3.

We can also extract the loop coordinates from a loop object using the methods a, b, and ab:

```
>> u = loop([-1 1 -2 0 -1 0]);
>> u.a

ans = -1     1     -2

>> u.b

ans =  0     -1     0

>> [a,b] = u.ab

a = -1     1     -2
b =  0     -1     0
```

As for braids, u.n returns the number of punctures (or strings).

9.3.1 Acting on Loops with Braids

Now we can act with braids on the loop u defined in the previous section. For example, we define the braid b to be σ_1^{-1} with 5 strings, corresponding to the 5 punctures, and then act on the loop 1 by using the multiplication operator:

```
>> u = loop([-1 1 -2 0 -1 0]);
>> b = braid([-1],5);   % one generator with 5 strings
>> b*u                  % act on a loop with a braid

ans = (( -1  1 -2  1 -1  0 ))
```

Figure 8.4 (right) shows the loop b*u. The first and second punctures were interchanged counterclockwise (the action of σ_1^{-1}), dragging the loop along.

The minimum length of an equivalence class of loops is determined by assuming the punctures are one unit of length apart and have zero size. After pulling tight the loop on the punctures, it is then made up of unit-length segments. The minimum length is thus an integer. For the loop in Fig. 8.2,

```
>> minlength(u)

ans = 12
```

Another useful measure of a loop's complexity is its minimum intersection number with the real axis [66, 88, 117], which for this loop is the same as its minimum length:

```
>> intaxis(u)

ans = 12
```

The `intaxis` method can be used to measure a braid's geometric complexity, as defined by Dynnikov and Wiest [44].

The `entropy` method of the `braid` class (Sect. 9.2) computes the topological entropy of a braid by repeatedly acting on a loop, and monitoring the growth rate of the loop. For example, let us compare the entropy obtained by acting 100 times on an initial loop, compared with the `entropy` method:

```
>> b = braid([1 2 3 -4]);
% apply braid 100 times to u, then compute growth of length
>> log(minlength(b^100*u)/minlength(u)) / 100

ans = 0.7637

>> entropy(b)

ans = 0.7672
```

The entropy value returned by `entropy(b)` is more precise, since that method monitors convergence and adjusts the number of iterations accordingly.

9.3.2 Loop Coordinates for a Braid

The command `loop(n,'BasePoint')` returns a *canonical set of loops* for n punctures:

```
>> u = loop(5,'BP')      % 'BP' is short for 'BasePoint'

ans = (( 0  0  0  0 -1 -1 -1 -1 ))*
```

This multiloop is depicted in Fig. 8.7, with basepoint puncture shown as a hollow dot on the right. The * indicates that this loop has a basepoint: the multiloop returned by `loop(5,'BP')` actually has 6 punctures, with the rightmost puncture meant to represent the boundary of a disk, or a base point for the fundamental group on a sphere with n punctures. The loops form a (non-oriented) generating set for the fundamental group of the disk with n punctures. The extra puncture thus plays no role dynamically, and `1.n` returns 5. The true total number of punctures, including the base point, is returned by `1.totaln`.

The canonical set of loops allows us to define loop coordinates for a braid, which is a unique normal form. The canonical loop coordinates for braids exploit the fact that two braids are equal if and only if they act the same way on the fundamental group of the disk (see [38] and Sects. 8.5 and 9.3.2). Hence, if we take a braid and act on `loop(5,'BP')`,

```
>> b = braid([1 2 3 -4]);
>> b*loop(5,'BP')

ans = (( 0  0  3 -1 -1 -1 -4  3 ))*
```

then the set of numbers `(( 0 0 3 -1 -1 -1 -4 3 ))*` can be thought of as *uniquely* characterizing the braid. It is this property that is used by braidlab to rapidly determine equality of braids. The same loop coordinates for the braid can be obtained without creating an intermediate loop with

```
>> loopcoords(b)

ans = (( 0  0  3 -1 -1 -1 -4  3 ))*
```

9.4 Entropy and Train Tracks

9.4.1 Topological Entropy and Complexity

There are a few methods that exploit the connection between braids and homeomorphisms of the punctured disk, as described in Chap. 4. Braids label *isotopy classes* of homeomorphisms, so we can assign a topological entropy to a braid:

```
>> entropy(braid([1 2 -3]))

ans = 0.8314
```

The entropy is computed by iterated action on a loop [88], as described in Sect. 8.4.3. This can fail if the braid is finite-order or has very low entropy:

```
>> entropy(braid([1 2]))
Warning: Failed to converge to requested tolerance; braid is
         likely finite-order or has low entropy.
         Returning zero entropy.

ans = 0
```

To force the entropy to be computed using the Bestvina–Handel train track algorithm [12], we add an optional `'Method'` parameter:

```
>> entropy(braid([1 2]),'Method','train')

ans = 0
```

Note that for large braids, the Bestvina–Handel algorithm is impractical.

The topological entropy is a measure of braid complexity that relies on iterating the braid. It gives the maximum growth rate of a "rubber band" anchored on the braid, as the rubber band slides up many repeated copies of the braid. For finite-order braids, this will converge to zero. The *geometric complexity* of a braid [44] is defined in terms of the $\log_2$ of the number of intersections of a set of curves with the real axis, after one application of the braid:

```
>> complexity(braid([1 -2]))

ans = 1.3863

>> complexity(braid([1 2]))

ans = 1.0986
```

See Sect. 9.3 or "`help braid.complexity`" for details on how the geometric complexity is computed.

9.4.2 Train Track Map and Transition Matrix

The Bestvina–Handel train track algorithm [12] (Chap. 7) can be used to determine the Thurston–Nielsen type of the braid as well as the train track map and its transition matrix [24, 33, 46, 125]:

```
>> train(braid([1 2 -3]))

ans = struct with fields:

         braid: [1x1 braidlab.braid]
        tntype: 'pseudo-Anosov'
       entropy: 0.8314
      transmat: [4x4 double]
         ttmap: {8x1 cell}

>> train(braid([1 2]))

ans = struct with fields:

         braid: [1x1 braidlab.braid]
        tntype: 'finite-order'
       entropy: 0
      transmat: [3x3 double]
         ttmap: {6x1 cell}
```

```
>> train(braid([1 2],4))   % reducing curve around 1,2,3

ans = struct with fields:

        braid: [1x1 braidlab.braid]
       tntype: 'reducible'
      entropy: 0
     transmat: [3x3 double]
        ttmap: {7x1 cell}
```

Braidlab uses Toby Hall's implementation of the Bestvina–Handel algorithm [64].

The train track map can be displayed in a human-readable format using the command ttmap:

```
>> tt = train(braid([1 2 -3]));
>> ttmap(tt)

 1 -> 4
 2 -> 1
 3 -> 2
 4 -> 3
 a -> D
 b -> d a -3 b -4 B
 c -> B 3 A
 d -> c
```

Here peripheral (infinitesimal) edges are denoted by numbers and main edges by letters. Inverse main edges are denoted by capital letters. The display of infinitesimal edges can be suppressed:

```
>> ttmap(tt,'Peripheral',false)

 a -> D
 b -> d a b
 c -> B A
 d -> c
```

The transition matrix associated with the train track map does *not* contain the peripheral edges, since these do not affect the entropy:

```
>> tt.transmat

ans =  0    1    1    0
       0    2    1    0
       0    0    0    1
       1    1    0    0

>> max(abs(eig(ans)))

ans = 2.2966
```

Up to a relabeling of the main edges, this is the same matrix (7.8) as we derived in Sect. 7.2.

9.5 Summary

In this chapter we introduced braidlab, a suite of Matlab functions for manipulating braids and loops. The main building blocks of braidlab are the classes `braid` and `loop`, which are both defined for punctured disks. These objects can be manipulated in a natural way: braids can be multiplied by other braids, and the multiplication of a braid and a loop denotes the Dynnikov action described in Sect. 8.2. Braidlab can efficiently compute the topological entropy of a braid, either by iterating its action on loops or using the Bestvina–Handel train track algorithm. The latter yields a lot more information about the mapping class labeled by the braid than its entropy, such as its train track map. However, the Bestvina–Handel algorithm can become prohibitively expensive to apply to large braids.

The viewpoint in this chapter was essentially algebraic: braids were described by a sequence of generators, with no reference to a geometric braid. In the next chapter we will see how to use braidlab to convert geometric braids (i.e., trajectories of particles) into algebraic braids.

Chapter 10
Braids and Data Analysis

> Speaker: *I think this will require more data.*
> Elderly professor: *I think this will require more thinking.*
>   (overheard at a seminar)

In the previous chapter we introduced the basic functionality of the software library braidlab. We saw how to create a braid by specifying the word in terms of Artin generators. In order to analyze the dynamics of real systems, we have to convert two-dimensional orbit data to braids. For systems with intrinsic periodicity, such as taffy pullers and many stirring devices, this is well-defined and leads to mathematically precise statements. For more complex real-world data, such as that arising from float trajectories in the ocean, the conversion is not exact and only makes sense in an asymptotic context. In this chapter we shall give examples of both cases.

10.1 Braids from Closed Trajectories

In Chap. 9 we regarded braids as geometrical objects and showed how to turn them into algebraic objects by writing them in terms of generators. But in practice how do we create braids from particle orbits? That is, if we have two-dimensional continuous trajectory data arising from some dynamical system, how do we turn this data into a braid? We will describe a simple procedure to do this in Sect. 10.1.1 and then give an example using taffy pullers in Sect. 10.1.2.

10.1.1 Constructing a Braid from Orbit Data

One simple technique for constructing a braid from two-dimensional continuous orbit data was described in Thiffeault [116, 117] and is as follows. We define an arbitrary line in the plane, called the *projection line*, and look at the ordering of particles projected along that line. We label each particle according to its order when projected on the line. In Fig. 10.1, for example, we see three particles labeled i, $i+1$, and $i+2$, and the projection line is the horizontal (dotted line).

© The Author(s), under exclusive license to Springer Nature Switzerland AG 2022
J.-L. Thiffeault, *Braids and Dynamics*, Frontiers in Applied Dynamical Systems:
Reviews and Tutorials 9, https://doi.org/10.1007/978-3-031-04790-9_10

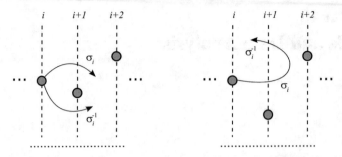

Fig. 10.1 Left: the two types of generators that occur when particles exchange their order along the projection line (dotted line at bottom). Right: a particle exchanging its order twice in a row with another leads to the generators σ_i followed by σ_i^{-1}, which cancel each other [After Thiffeault [116, 117].]

The position of the particles evolves with time. A *crossing* occurs whenever any two particles exchange their order along the projection line. (Only adjacent particles can exchange their order at a given time, since the trajectories are continuous.) For particle i exchanging its order with particle $i+1$, we assign a generator σ_i to the crossing if the particle on the left passes above the one on the right and σ_i^{-1} if it passes below (Fig. 10.1, left). As we watch the trajectories unfold, we construct an algebraic braid as a sequence of generators, with each generator corresponding to a crossing, and their order determined by when each crossing occurs. Note that when two particles exchange order along the crossing line twice as in Fig. 10.1 (right), without exchanging positions vertically, then the two crossings yield the generators σ_i followed by σ_i^{-1}, which cancel.

If the trajectories end up at the same set of spatial points as they started, then the outcome of this procedure is a true geometric braid.

10.1.2 An Example: Taffy Pullers

Taffy pullers are a class of devices designed to stretch and fold soft candy repeatedly [50, 118], which we introduced in Chap. 1. The goal of the repeated folding is to aerate the taffy. Since many such folds are required, the process has been mechanized using devices with a combination of fixed and moving rods. The two most typical designs are shown in Fig. 10.2: the one on the left has a single fixed rod (gray) and two moving rods, each rotating on a different axis. The one on the right in Fig. 10.2 has four moving rods, sharing two axes of rotation. See Fig. 1.2 on page 3 for a picture of a three-rod taffy puller.

Let us use braidlab to analyze the taffy puller rod motion. First we will define the trajectories of the three rods in Fig. 10.2 (left). We define an angle to serve as a time parameter:

```
>> npts = 200;
```

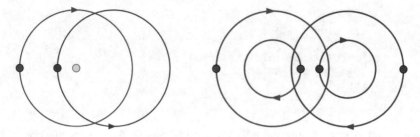

Fig. 10.2 Left: three-rod taffy puller. Right: four-rod taffy puller

```
>> t = linspace(-pi,pi,npts);
```

The angle parameter t consists of a vector of 200 points spanning $[-\pi, \pi]$. Now we define the three-rod orbits in the complex plane:

```
>> n = 3;                          % the number of rods
>> Z = zeros(npts,n);              % allocate array
>> Z(:,1) = 0 + .75*exp(-1i*t);    % red orbit centered at 0
>> Z(:,2) = 0;                     % gray orbit fixed rod at 0
>> Z(:,3) = .5 + .75*exp(1i*t);    % blue orbit centered at .5
```

The complex exponentials trace out the circular orbits in the complex plane. We then convert these complex orbits to a braid with the constructor:

```
>> b = braid(Z)

b = < -2  1  1 -2 >
```

We already discussed these orbits and the associated braid in Sect. 4.2. The three orbits are plotted as a space-time diagram in Fig. 4.2a, and their standard braid diagram is shown in Fig. 4.2b. Braidlab has automatically determined that the algebraic braid is $\sigma_2^{-1}\sigma_1^2\sigma_2^{-1}$, using essentially the projection method described above. By default, the projection line is taken to be the real axis.

For convenience, the taffy command computes braids for standard taffy pullers by tracing the rod orbits and reconstructing the braid as described above. From the folder doc/examples, run the command

```
>> b = taffy('3rods')

b = < -2  1  1 -2 >
```

which also produces Fig. 3.8.

Once we have transformed the orbits into a sequence of braid generators, we can use braidlab to compute some useful quantities. For instance, the Thurston–Nielsen type and topological entropy of this braid are easily obtained by constructing the train track:

```
>> train(b)
```

```
ans = struct with fields:

      braid: [1x1 braidlab.braid]
     tntype: 'pseudo-Anosov'
    entropy: 1.7627
   transmat: [2x2 double]
      ttmap: {5x1 cell}
```

One would expect a competent taffy puller to be pseudo-Anosov, as this one is. It implies that there is no "bad" initial condition where a piece of taffy never gets stretched or stretches slowly. A reducible or finite-order braid would indicate poor design.

The entropy is a measure of the taffy puller's effectiveness: it gives the rate of growth of curves anchored on the rods. Thus, the length of the taffy is multiplied (asymptotically) by $e^{1.7627} \simeq 5.828$ for each full period of rod motion. Needless to say, this leads to extremely rapid growth, since after 10 periods the taffy length has been multiplied by roughly 10^7.

The four-rod design on the right in Fig. 10.2 can be plotted and analyzed with

```
>> b = taffy('4rods')

b = < 1  3  2  2  1  3 >
```

When we apply `train` to this braid, we find

```
>> train(b)

ans =

  struct with fields:

        braid: [1x1 braidlab.braid]
       tntype: 'pseudo-Anosov'
      entropy: 1.7627
     transmat: [3x3 double]
        ttmap: {7x1 cell}
```

Again, the braid is pseudo-Anosov with exactly the same entropy as the 3-rod taffy puller, 1.7627. There is thus no obvious advantage to using more rods in this case. Historically, it is not clear if the 4-rod taffy puller was thought to be better than the 3-rod design, but mathematically, they are identical. See Thiffeault [118] for more details.

Can we improve the 4-rod taffy puller so that it *does* perform better than the 3-rod design? Braidlab allows us to experiment quickly with different prototypes. A simple modification of the 4-rod design in Fig. 10.2 is shown on the left in Fig. 10.3. The only change is that we extended the rotation axles into two extra fixed rods (shown in gray). The resulting braid is

```
>> b = taffy('6rods-bad')

b = < 2  1  2  4  5  4  3  3  2  1  2  4  5  4 >
```

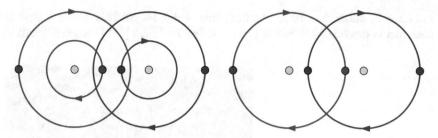

Fig. 10.3 Left: a six-rod taffy puller based on Fig. 10.2, with two added fixed rods (gray). This is a poor design, since it leads to a reducible braid. Right: the same as on the left, but with the same radius of motion for all the moving rods. The braid is in this case pseudo-Anosov, with larger entropy than the 4-rod design

with Thurston–Nielsen type

```
>> train(b)

ans = struct with fields:

        braid: [1x1 braidlab.braid]
       tntype: 'reducible'
      entropy: 0
     transmat: [5x5 double]
        ttmap: {11x1 cell}
```

There are reducing curves in this design: simply wrap a loop around the left gray rod and the inner red rod, and it will rotate without stretching. Reducing curves are undesirable in taffy puller designs since they indicate redundant rods. To avoid this, we extend the radius of motion of the inner rods to equal that of the outer ones and obtain the design shown in Fig. 10.3 (right). The corresponding braid is

```
>> b = taffy('6rods')

b = < 3  2  1  2  4  5  4  3  3  2  1  2  5  4  5  3 >
```

with the Thurston–Nielsen type and entropy

```
>> t = train(b)

ans = struct with fields:

        braid: [1x1 braidlab.braid]
       tntype: 'pseudo-Anosov'
      entropy: 2.6339
     transmat: [5x5 double]
        ttmap: {11x1 cell}
```

The fixed rods have increased the entropy by 50%! This sounds like a fairly small change, but what it means is that this 6-rod design achieves growth of 10^7 in about 6 full periods, rather than 10 for the traditional 4-rod design. This device is easy to construct in practice, as shown in Fig. 10.4. See Thiffeault [118] for more details.

Fig. 10.4 A device realizing the rod orbits in Fig. 10.3 (right), built by the author and Alex Flanagan

We used taffy pullers as an illustration, but there are many other real-world systems where closed braids occur naturally. For instance, because of practical considerations, many rod-stirring devices are time-periodic, so their properties are readily analyzed using braids [15, 16, 27, 50, 51, 52, 53, 60, 111, 127, 128, 129]. Some systems also have distinguished periodic orbits whose braiding motion can be studied, such as some configurations of point vortices [28] and defects in active nematics [107, 114].

10.1.3 Changing the Projection Line

The symmetric designArtin generators of the taffy pullers illustrates one pitfall when constructing braids. If we give an optional projection line angle of $\pi/2$ to taffy, we get an error:

```
>> taffy('4rods',pi/2)
Error using braidlab.braid/colorbraiding
Paths of particles 2 and 1 have a coincident projection.
Try changing the projection angle.
```

This corresponds to using the y (vertical) axis to compute the braid, but we can see from Fig. 10.2 that this is a bad choice, since all the rods are initially perfectly aligned along that axis. The braid obtained would depend sensitively on numerical roundoff when comparing the rod projections. Instead of attempting to construct the braid, braidlab returns an error and asks the user to modify the projection line. A tiny change in the projection line is sufficient to break the symmetry:

```
>> taffy('4rods',pi/2 + .01)
```

```
ans = < -2  2  1  3  2 -3 -1  3  1  2  1  3 >

>> compact(ans)

ans = < 3  1  2  2  3  1 >
```

which is actually equal to the braid formed from projecting on the x axis, though it need only be conjugate (see Sect. 10.2.2).

10.2 Braids from Non-closed Trajectories

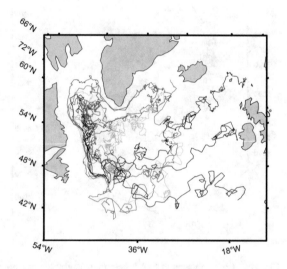

Fig. 10.5 Twelve float trajectories in the Labrador sea. From [101, 117]

In this section we leave the comforting embrace of mathematical rigor. When dealing with real-world data, such as the particle trajectories of Figs. 10.5 and 10.6, the trajectories are not in general closed: the final position of the particles does not correspond to their initial position, even as a set. Nevertheless, the trajectories still "entangle" in some sense and the braid approach can help us in quantifying that entanglement.

10.2.1 Constructing a Braid from Data: An Example

We can assign a braid to trajectory data by looking for crossings along a projection line as in Sect. 10.1.1). The braid constructor allows us to do this easily. The folder `testsuite/testcases` contains a dataset of trajectories, from laboratory data for granular media [100]. We load the data:

```
>> clear; load testdata
>> whos
  Name          Size                    Bytes  Class     Attributes

  XY           9739x2x4                623296  double
  ti              1x9739                77912  double
```

Here `ti` is the vector of times, and `XY` is a three-dimensional array: its first component specifies the timestep, its second specifies the X or Y coordinate, and its third specifies one of the 4 particles. Figure 10.6 (left) shows the X and Y coordinates

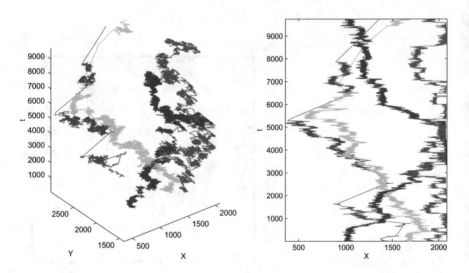

Fig. 10.6 Left: a dataset of four trajectories. Right: the trajectories projected along the X axis

of these four trajectories, with time plotted vertically. Figure 10.6 (right) shows the same data, but projected along the X direction. To construct a braid from this data, we simply execute

```
>> warning('off','BRAIDLAB:braid:colorbraiding:notclosed')
>> b = braid(XY);
>> b.length

ans = 894
```

This is a very long braid! (We have temporarily turned off a warning about the data not being closed; we will explain this in Sect. 10.2.2 below.) But Fig. 10.6 suggests that this is misleading: many of the crossings are "wiggles" that cancel each other out. Indeed, if we attempt to shorten the braid,

```
>> b = compact(b)
```

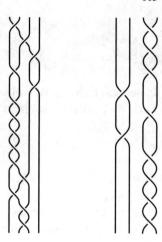

Fig. 10.7 Left: the compacted braid $\sigma_2^{-2}\sigma_1^{-1}\sigma_2^{-1}\sigma_1^{-5}\sigma_3\sigma_1^{-1}\sigma_3\sigma_2\sigma_1$ corresponding to the X projection of the data in Fig. 10.6, with closure enforced. Right: the compacted braid $\sigma_3^{-4}\sigma_1\sigma_3^{-1}\sigma_1\sigma_3^{-3}$ corresponding to the Y projection. The two braids are conjugate

```
b = < -2 -2 -1 -2 -1 -1 -1 -1 -1  3 -1  3  2  1 >

>> b.length

ans = 14
```

we find the number of generators (the length) has dropped to 14! We can then plot this shortened braid as a braid diagram using `plot(b)` to produce Fig. 10.7. The braid diagram allows us to see some topological information clearly, such as the fact that the second and third particles undergo a large number of twists around each other; we can check this by creating a subbraid with only those two strings:

```
>> subbraid(b,[2 3])

ans = < -1 -1 -1 -1 -1 -1 -1 -1 >
```

which shows that the winding number between these two strings is -4: each of the eight `-1`s corresponds to a half-twist of the two trajectory strings.

10.2.2 Changing the Projection Line and Enforcing Closure

The braid in the previous section was constructed from the data by assuming a projection along the X axis (the default). We can choose a different projection by specifying an optional angle for the projection line; for instance, to project along the Y axis, we invoke

```
>> b = braid(XY,pi/2);   % project onto Y axis
>> b.length

ans = 673
```

```
>> b.compact

ans = < -3 -3 -3 -3  1 -3 -3 -3 -3 >
```

In general, a change of projection line only changes the braid by conjugation [24, 117]. We can test for conjugacy:

```
>> bX = compact(braid(XY,0)); bY = compact(braid(XY,pi/2));
>> conjtest(bX,bY)    % test for conjugacy of braids

ans = 0
```

The braids are not conjugate. This is because our trajectories do not form a "true" braid: the final points do not correspond exactly with the initial points, as a set. This is the reason why we turned off a warning in Sect. 10.2.1: braidlab warns us when we are trying to create a braid from data that does not "join up."

If we truly want a rotationally conjugate braid out of our data, we need to enforce a closure method:

```
>> XY = closure(XY);    % close braid and avoid new crossings
>> bX = compact(braid(XY,0)), bY = compact(braid(XY,pi/2))

bX = < -2 -2 -1 -2 -1 -1 -1 -1 -1  3 -1  3  2  1 >

bY = < -3 -3 -3 -3  1 -3  1 -3 -3 -3 >
```

This default closure simply interpolates line segments from the final points to the initial points in such a way that no new crossings are created in the X projection. Hence, the X-projected braid bX is unchanged by the closure, but here the Y-projected braid bY is longer by one generator (bY is plotted on the right in Fig. 10.7). This is enough to make the braids conjugate:

```
>> [~,c] = conjtest(bX,bY)    % ~ means discard first return arg

c = < 3  2 >
```

where the optional second argument c is the conjugating braid, as we can verify:

```
>> bX == c*bY*c^-1

ans = 1
```

There are other ways to enforce closure of a braid (see help closure), in particular closure(XY,'MinDist'), which minimizes the total distance between the initial and final points. However, none of the methods is ideal and enforcing closure should be done carefully, since it can easily introduce spurious entropy in the resulting braid. In general, when dealing with data, it is better not to close the braid at all, and instead use the databraid class, described in Sect. 10.2.3 below.

Note that conjtest uses the library *CBraid* [35] to first convert the braids to Garside canonical form [19] and then to determine conjugacy. This is very inefficient, so is impractical for large braids.

10.2.3 Finite-Time Braiding Exponent (FTBE)

We mentioned at the start of this section that one of the motivations for forming braids out of trajectory data is to quantify the degree of entanglement of the trajectories. The degree of entanglement can serve as a proxy measure of mixing in fluid dynamics. The reasoning is that the braid formed by the trajectories captures the topological entropy of the underlying flow [60, 127, 128, 129]. This approach has been used to quantify mixing in oceanic float trajectories [117], laboratory experiments [39, 48], and model mixing systems [3, 29, 32, 56, 60, 115, 116, 132]. Quantifying the entanglement has also proved useful even when there is no underlying flow, such as for studying the motion of particles in a granular medium [100], agents in crowd dynamics [2, 84], and flocks of birds [34].

Topological entropy by itself is not well-suited to describing the degree of entanglement. The method used for its computation in Chap. 8 is based on repeatedly acting on a loop with a given braid. When the braid is obtained from trajectory data, it is generally not a good idea to iterate the braid action in this manner: the underlying trajectories are not closed, and iteration requires enforcing a closure condition (Sect. 10.2.2). The closure method itself may introduce spurious entropy. The remedy to this is to (hopefully) have enough data that there is no need to iterate the braid. This will not be possible in every circumstance, but it is the best we can hope for.

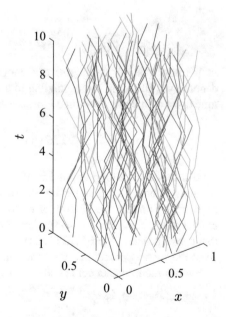

Fig. 10.8 A braid of 50 random walkers on the unit square, each taking 10 steps

We illustrate the approach using randomized trajectories. Braidlab provides a command `randomwalk` to generate trajectory data corresponding to independent random walkers. Each random walker takes a small step in a uniformly chosen direction inside the unit square, reflecting at the boundaries as needed. The commands

```
>> rng('default')
>> XY = randomwalk(50,10,.1); t = 0:10;
```

generate trajectory data for 50 random walkers, each taking 10 random step of length 0.1, as depicted in Fig. 10.8. The variable t is the time for each step. We construct a braid from this trajectory data using

```
>> B = databraid(XY,t);
```

A databraid object is very similar to a braid, except that it does not require the data to be closed. In addition, the time coordinate t is passed as an argument, since the data need not be sampled at equal time intervals. The databraid object contains the algebraic braid traced out by the trajectories:

```
>> B.braid

ans = < 27   26  -11  -27   10  -16   49    8   -6  37   19   48  -41 ... >
>> length(B.braid)

ans = 930
```

In addition, the databraid object records the times at which crossings occur along the projection line:

```
>> B.tcross

ans = 0.0018   0.0040   0.0084   0.0317   0.0419   0.0574   0.0867  ...
```

We now quantify the degree of entanglement of our random trajectories. If we denote by γ_t the braid corresponding to a collection of n trajectories sampled up to time t, we define the *finite-time braiding exponent* (FTBE)[1] as

$$\text{FTBE}(\gamma_t) = \frac{1}{t} \log \frac{L(u \cdot \gamma_t)}{L(u)} . \tag{10.1}$$

Here u is the multiloop defined in Eq. (8.48), and $L(u)$ is the minimum number of intersections (8.8). The definition (10.1) is very similar in spirit to the iterative approach to finding the dilation or entropy of a braid, as given by Eq. (8.46). The crucial difference is that there is no iteration: we rely instead on t being large enough that convergence of the FTBE is observed in practice. In that sense the FTBE measures the "entanglement per unit time" of a collection of trajectories.

For our random data defined above, we find the FTBE with

```
>> ftbe(B)

ans = 0.5359
```

---

[1] The name FTBE comes from the analogy with finite-time Lyapunov exponents.

The FTBE is a sensible measure of the rate entanglement when it has converged in time. We can test for this convergence by generating longer and longer trajectories, and computing the FTBE as a function of time. The result of doing this is shown

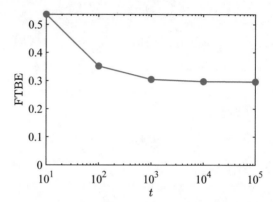

Fig. 10.9 The finite-time braiding exponent (FTBE) as a function of trajectory length for a random walk of 50 particles in the unit square

in Fig. 10.9. Note the convergence for large t: the final FTBE value is about 0.3 for $t = 10^5$. For this large time, the 50 particles have undergone about 10^7 crossings. See Budišić and Thiffeault [29] and Filippi et al. [48] for more details on FTBEs and their use.

10.3 Summary

Algebraic braids are an elegant way of encoding topological information contained in two-dimensional trajectory data. The conversion of the data to a braid is simple, and can be done rapidly with braidlab. When the trajectories are "naturally" closed, such as for periodic devices like taffy pullers, then the result is a true braid. The braid can then be used to deduce information about the device, such as whether the rod motion labels a pseudo-Anosov mapping class, or its induced topological entropy.

When the trajectory data is not naturally closed, as is the case in many real-world situations, then the conversion to a braid is less well-defined mathematically, though it has proven extremely convenient and useful. In that case, it is crucial to have long trajectories, so that the FTBE (the finite-tie braiding exponent, a running measure of topological entropy) has enough time to converge.

There are many remaining challenges in using braids for data analysis. A solid foundation for "open" braids is lacking, as is a complete description of what dynamical information can be extracted from them. Allshouse and Thiffeault [3] and Yeung et al. [132], for example, have used braids to quantify coherence of sets, which is an important aspect of understanding transport. There are also non-braid-based topological approaches that show promise, especially if they can be generalized to apply to three-dimensional trajectory data [103, 114].

Appendix A
Derivation of Dynnikov Update Rules (Spencer A. Smith)

We provide a derivation of the Dynnikov update rules for the braid generators, as given in Sect. 8.2. The derivation is based on a single Whitehead move that involves switching the shared edge of two adjacent triangles in a triangulation.

A.1 Dynnikov Coordinates

Recall that in Fig. 8.3 (p. 81) we defined the intersection numbers v_i, $0 \le i \le n$, and μ_i, $1 \le i \le 2n-4$, for a loop on a disk with n punctures. These numbers count transverse intersections of a given loop with the μ_i and v_i curves. (We use μ_i and v_i to denote the intersection numbers as well as the curves.) We can identify the disk (including punctures and μ_i & v_i curves) with a triangulation on the 2-sphere with $n+1$ punctures. We are including the boundary of the disk as one of these points.

Given a loop (or more generally, a measured lamination), we have a coordinate representation in terms of the μ_i and v_i variables. Conversely, if we have the coordinates μ_i and v_i (which obey triangle inequalities for each triangle in the triangulation), we can draw the loop it represents (up to isotopy). When encoding a loop, we assume that the loop has been pulled tight so that we cannot reduce the number of intersections with μ_i and v_i by sliding the loop around; this precludes scenarios where the loop enters a triangle and immediately leaves via the same edge. This condition is maintained by the update rules.

Recall the definition (8.2) for the differences in intersection numbers,

$$a_i = \tfrac{1}{2}(\mu_{2i} - \mu_{2i-1}), \qquad b_i = \tfrac{1}{2}(v_i - v_{i+1}), \tag{A.1}$$

for $i = 1, \ldots, n-2$. The isotopy class of a loop is then denoted by a Dynnikov coordinate vector $u = (a_1, \cdots, a_{n-2}, b_1, \cdots, b_{n-2})$.

This appendix is based on unpublished notes by Spencer A. Smith and was edited by J-LT to harmonize with the notation in the rest of the book. See also Smith and Dunn [108] for an abridged description.

© The Author(s), under exclusive license to Springer Nature Switzerland AG 2022
J.-L. Thiffeault, *Braids and Dynamics*, Frontiers in Applied Dynamical Systems:
Reviews and Tutorials 9, https://doi.org/10.1007/978-3-031-04790-9

The Dynnikov update rules tell us how the Dynnikov coordinates are modified by a generator $\sigma_i \in B_n$, $i = 1, \ldots, n-1$, as detailed in Sect. 8.2. For instance, the update rules for σ_i^{-1}—the counterclockwise (CCW) exchange of punctures i and $i+1$—were given by (8.18)–(8.19) as

$$\bar{a}_{i-1} = a_{i-1} + b_{i-1}^+ + \left(b_i^+ - d_{i-1}\right)^+, \tag{A.2a}$$

$$\bar{b}_{i-1} = b_i - d_{i-1}^+, \tag{A.2b}$$

$$\bar{a}_i = a_i + b_i^- + \left(b_{i-1}^- + d_{i-1}\right)^-, \tag{A.2c}$$

$$\bar{b}_i = b_{i-1} + d_{i-1}^+, \tag{A.2d}$$

for $i = 2, \ldots, n-2$, with the definition

$$d_{i-1} = a_{i-1} - a_i + b_i^+ - b_{i-1}^-. \tag{A.3}$$

We used the operators $f^+ := \max(f, 0)$ and $f^- := \min(f, 0)$. The overbar denotes the new Dynnikov coordinates, after the exchange. See Fig. 8.4 for an illustration.

A.2 Whitehead Moves

A.2.1 Triangulation Coordinates

As mentioned before, the disk along with the μ_i and ν_i curves can be thought of as a triangulation of the 2-sphere (with $n+1$ points). The Euler characteristic is then $\chi = 2 = F - E + V = F - E + (n+1)$, with the standard face–edge–vertex notation. Also, since there are 2 faces adjacent to each edge and 3 edges adjacent to each face, we have $3F = 2E$. Combining, we have $E = 3n - 3$ for any triangulation of the sphere with $n+1$ points.

A.2.2 Whitehead Move and Update Rule

We can maintain any triangulation as the vertices move by executing a series of static local re-triangulations (called Whitehead moves) and regular point movement. The Whitehead moves update the coordinates associated with the edges of the triangulation. A Whitehead move is shown in Fig. A.1, as is an example.

Only one edge (E) needs updating for a Whitehead move, and the update rule is particularly simple:

$$E' = \max(A + C, B + D) - E. \tag{A.4}$$

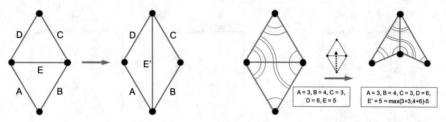

Fig. A.1 A single Whitehead move on a pair of adjacent triangles removes the shared edge, and it replaces it with an edge connecting the other two vertices of the two triangles. The righthand image shows an example

All of the Dynnikov coordinate update rules can be derived from this one rule. It will be useful to rewrite this in the max-plus algebra notation (Sect. 8.3), where $[\![x+y]\!] \equiv \max(x,y)$ and $[\![xy]\!] \equiv x+y$. The double brackets remind us that additions and multiplications inside the brackets are to be interpreted in the max-plus sense. The Whitehead update rule (A.4) can be written as

$$E' = \left[\!\!\left[\frac{AC+BD}{E}\right]\!\!\right]. \tag{A.5}$$

A.3 Deriving the Update Rules

We now give a derivation of all the Dynnikov update rules, using the Whitehead update rule (A.5).

A.3.1 Counterclockwise Switch

Consider the counterclockwise (CCW) interchange of points i and $i+1$, encoded as Artin braid generator σ_i^{-1}. This motion is depicted in Fig. A.2 as decomposed into a series of Whitehead moves and physical movement of the points. Moving from A to B, we see that

$$v'_i = \max(\mu_{2i-2}+\mu_{2i-1}, \mu_{2i-3}+\mu_{2i}) - v_i.$$

The points physically move between B and C (red arrows), but the triangulation has not changed topologically. Between C and D, there are two Whitehead moves, and we get

$$\mu'_{2i-3} = \max(\mu_{2i-2}+\mu_{2i-1}, v'_i+v_{i-1}) - \mu_{2i-3}$$

and

$$\mu'_{2i} = \max(\mu_{2i-2}+\mu_{2i-1}, v'_i+v_{i+1}) - \mu_{2i}.$$

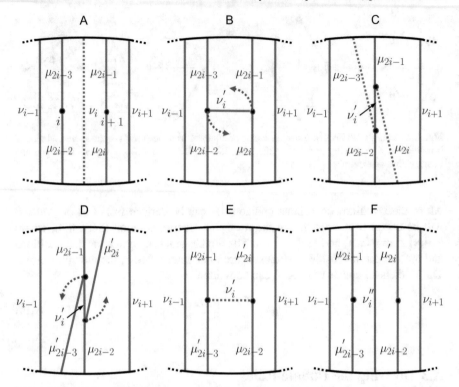

Fig. A.2 The series of Whitehead moves and physical movements of the two points involved in the counterclockwise braid generator σ_i^{-1}. The dashed blue lines denote an immanent Whitehead move, and the dashed red arrows indicate physical movement of the points

From D to E, the points again move, maintaining the triangulation topology. Finally, from E to F, we have the last Whitehead move and get

$$v_i'' = \max(\mu_{2i-2} + \mu_{2i-1}, \mu_{2i}' + \mu_{2i-3}') - v_i'.$$

Altogether, in max-plus notation:

$$v_i' = \left[\!\!\left[\frac{\mu_{2i-2}\mu_{2i-1} + \mu_{2i-3}\mu_{2i}}{v_i}\right]\!\!\right], \qquad \mu_{2i-3}' = \left[\!\!\left[\frac{\mu_{2i-2}\mu_{2i-1} + v_i'v_{i-1}}{\mu_{2i-3}}\right]\!\!\right] \qquad \text{(A.6a)}$$

$$\mu_{2i}' = \left[\!\!\left[\frac{\mu_{2i-2}\mu_{2i-1} + v_i'v_{i+1}}{\mu_{2i}}\right]\!\!\right], \qquad v_i'' = \left[\!\!\left[\frac{\mu_{2i-2}\mu_{2i-1} + \mu_{2i}'\mu_{2i-3}'}{v_i'}\right]\!\!\right]. \qquad \text{(A.6b)}$$

We also have a permutation to consider. We want to identify (via position) the variables in Fig. A.2F with the updated (overbar) versions of those in Fig. A.2A. We will use the overbar (e.g., $\bar{v}$) to denote the coordinate after the braid generator action (the primes are used for the intermediate steps given by the Whitehead moves). Define

$$v = (\mu'_{2i-3}, \mu_{2i-2}, \mu_{2i-1}, \mu''_{2i}, v''_i),$$
$$\bar{v} = (\bar{\mu}_{2i-3}, \bar{\mu}_{2i-2}, \bar{\mu}_{2i-1}, \bar{\mu}_{2i}, \bar{v}_i).$$

Now our new coordinates are

$$\bar{v}_k = v_{\pi(k)} \quad \text{for} \quad \pi = (3,1,4,2,5),$$

where π is a permutation.

The Dynnikov coordinates that will need to be updated are

$$a_{i-1} = \frac{1}{2}\left[\!\left[\frac{\mu_{2i-2}}{\mu_{2i-3}}\right]\!\right], \quad a_i = \frac{1}{2}\left[\!\left[\frac{\mu_{2i}}{\mu_{2i-1}}\right]\!\right], \quad b_{i-1} = \frac{1}{2}\left[\!\left[\frac{v_{i-1}}{v_i}\right]\!\right], \quad b_i = \frac{1}{2}\left[\!\left[\frac{v_i}{v_{i+1}}\right]\!\right].$$

The updated versions are

$$\bar{a}_{i-1} = \frac{1}{2}\left[\!\left[\frac{\bar{\mu}_{2i-2}}{\bar{\mu}_{2i-3}}\right]\!\right], \quad \bar{a}_i = \frac{1}{2}\left[\!\left[\frac{\bar{\mu}_{2i}}{\bar{\mu}_{2i-1}}\right]\!\right], \quad \bar{b}_{i-1} = \frac{1}{2}\left[\!\left[\frac{\bar{v}_{i-1}}{\bar{v}_i}\right]\!\right], \quad \bar{b}_i = \frac{1}{2}\left[\!\left[\frac{\bar{v}_i}{\bar{v}_{i+1}}\right]\!\right].$$

Now we just plug in our results from the updated μ_i and v_i and simplify and write in terms of the Dynnikov a_i and b_i.

Let us start with $\bar{a}_{i-1}$:

$$\bar{a}_{i-1} = \frac{1}{2}\left[\!\left[\frac{\bar{\mu}_{2i-2}}{\bar{\mu}_{2i-3}}\right]\!\right] = \frac{1}{2}\left[\!\left[\frac{\mu'_{2i-3}}{\mu_{2i-1}}\right]\!\right] = \frac{1}{2}\left[\!\left[\frac{\mu_{2i-2}\mu_{2i-1} + v'_i v_{i-1}}{\mu_{2i-3}\mu_{2i-1}}\right]\!\right],$$

where we used (A.6a) for μ'_{2i-3}, and use it again for v'_i to get

$$\bar{a}_{i-1} = \frac{1}{2}\left[\!\left[\frac{\mu_{2i-2}}{\mu_{2i-3}} + \frac{(\mu_{2i-2}\mu_{2i-1} + \mu_{2i-3}\mu_{2i})v_{i-1}}{v_i\mu_{2i-3}\mu_{2i-1}}\right]\!\right]$$

$$= \frac{1}{2}\left[\!\left[\frac{\mu_{2i-2}}{\mu_{2i-3}} + \left(\frac{\mu_{2i-2}}{\mu_{2i-3}} + \frac{\mu_{2i}}{\mu_{2i-1}}\right)\frac{v_{i-1}}{v_i}\right]\!\right]$$

$$= \left[\!\left[a_{i-1} + (a_{i-1} + a_i)b_{i-1}\right]\!\right]$$

$$= \left[\!\left[a_{i-1}\left(1 + \left(1 + \frac{a_i}{a_{i-1}}\right)b_{i-1}\right)\right]\!\right]. \tag{A.7}$$

Dropping the max-plus algebra notation, we have

$$\bar{a}_{i-1} = \max(a_{i-1}, \max(a_{i-1}, a_i) + b_{i-1}) = a_{i-1} + ((a_i - a_{i-1})^+ + b_{i-1})^+.$$

Next, we find update rule for a_i:

$$\bar{a}_i = \frac{1}{2}\left[\!\left[\frac{\bar{\mu}_{2i}}{\bar{\mu}_{2i-1}}\right]\!\right] = \frac{1}{2}\left[\!\left[\frac{\mu_{2i-2}}{\mu'_{2i}}\right]\!\right] = \frac{1}{2}\left[\!\left[\frac{\mu_{2i-2}\mu_{2i}}{\mu_{2i-2}\mu_{2i-1} + v'_i v_{i+1}}\right]\!\right],$$

where we used (A.6b) for μ'_{2i}. We invert the interior of the double bracket and use (A.6a) for v'_i:

$$
\begin{aligned}
\bar{a}_i &= \tfrac{1}{2}\left[\!\left[\left(\frac{\mu_{2i-1}}{\mu_{2i}} + \frac{v'_i v_{i+1}}{\mu_{2i-2}\mu_{2i}}\right)^{-1}\right]\!\right] \\
&= \tfrac{1}{2}\left[\!\left[\left(\frac{\mu_{2i-1}}{\mu_{2i}} + \frac{(\mu_{2i-2}\mu_{2i-1} + \mu_{2i-3}\mu_{2i})v_{i+1}}{\mu_{2i-2}\mu_{2i}v_i}\right)^{-1}\right]\!\right] \\
&= \tfrac{1}{2}\left[\!\left[\left(\frac{\mu_{2i-1}}{\mu_{2i}} + \left(\frac{\mu_{2i-1}}{\mu_{2i}} + \frac{\mu_{2i-3}}{\mu_{2i-2}}\right)\frac{v_{i+1}}{v_i}\right)^{-1}\right]\!\right] \\
&= \left[\!\left[\left(\frac{1}{a_i}\left(1 + \left(1 + \frac{a_i}{a_{i-1}}\right)\frac{1}{b_i}\right)\right)^{-1}\right]\!\right].
\end{aligned}
$$ (A.8)

Dropping the max-plus algebra notation, we have finally

$$
\begin{aligned}
\bar{a}_i &= -\max(-a_i, \max(-a_i, -a_{i-1}) - b_i) \\
&= \min(a_i, \min(a_i, a_{i-1}) + b_i) \\
&= a_i - \left((a_i - a_{i-1})^+ - b_i\right)^+ \\
&= a_i + \left((a_{i-1} - a_i)^- + b_i\right)^-,
\end{aligned}
$$

where we used $-(f)^{\pm} = (-f)^{\mp}$.

Next we find the update rule for b_{i-1}:

$$
\bar{b}_{i-1} = \tfrac{1}{2}\left[\!\left[\frac{\bar{v}_{i-1}}{\bar{v}_i}\right]\!\right] = \tfrac{1}{2}\left[\!\left[\frac{v_{i-1}}{v''_i}\right]\!\right] = \tfrac{1}{2}\left[\!\left[\left(\frac{\mu_{2i-2}\mu_{2i-1} + \mu'_{2i}\mu'_{2i-3}}{v_{i-1}v'_i}\right)^{-1}\right]\!\right],
$$

where we used (A.6b) for v''_i. We use Eqs. (A.6) a few more times:

$$
\begin{aligned}
\bar{b}_{i-1} &= \tfrac{1}{2}\left[\!\left[\left(\frac{\mu_{2i-2}\mu_{2i-1}}{v_{i-1}v'_i} + \frac{\left(\mu_{2i-2}\mu_{2i-1} + v_{i+1}v'_i\right)\left(\mu_{2i-2}\mu_{2i-1} + v_{i-1}v'_i\right)}{\mu_{2i}\mu_{2i-3}v_{i-1}v'_i}\right)^{-1}\right]\!\right] \\
&= \tfrac{1}{2}\left[\!\left[\left(\frac{v_i}{v_{i-1}}\frac{1 + \frac{\mu_{2i-2}\mu_{2i-1}}{\mu_{2i-3}\mu_{2i}}\left(1 + \frac{v_{i+1}}{v_i}\left(1 + \frac{\mu_{2i-3}\mu_{2i}}{\mu_{2i-2}\mu_{2i-1}}\right)\right)\left(1 + \frac{v_{i-1}}{v_i}\left(1 + \frac{\mu_{2i-3}\mu_{2i}}{\mu_{2i-2}\mu_{2i-1}}\right)\right)}{\left(1 + \frac{\mu_{2i-3}\mu_{2i}}{\mu_{2i-2}\mu_{2i-1}}\right)}\right)^{-1}\right]\!\right] \\
&= \left[\!\left[\left(\frac{1 + \frac{a_{i-1}}{a_i}\left(1 + \frac{1}{b_i}\left(1 + \frac{a_i}{a_{i-1}}\right)\right)\left(1 + b_{i-1}\left(1 + \frac{a_i}{a_{i-1}}\right)\right)}{b_{i-1}\left(1 + \frac{a_i}{a_{i-1}}\right)}\right)^{-1}\right]\!\right] \\
&= \left[\!\left[\frac{b_{i-1}\left(1 + \frac{a_i}{a_{i-1}}\right)}{1 + \frac{a_{i-1}}{a_i}\left(1 + \frac{1}{b_i}\left(1 + \frac{a_i}{a_{i-1}}\right)\right)\left(1 + b_{i-1}\left(1 + \frac{a_i}{a_{i-1}}\right)\right)}\right]\!\right].
\end{aligned}
$$ (A.9)

Written in the $f^{\pm}$ notation, this is

$$\bar{b}_{i-1} = b_{i-1} + (a_i - a_{i-1})^+$$
$$- \left(\left((a_i - a_{i-1})^+ - b_i \right)^+ - (a_i - a_{i-1}) + \left((a_i - a_{i-1})^+ + b_{i-1} \right)^+ \right)^+. \quad (A.10)$$

Now for the final update rule, for b_i:

$$\bar{b}_i = \frac{1}{2} \left[\!\left[\frac{\bar{v}_i}{\bar{v}_{i+1}} \right]\!\right] = \frac{1}{2} \left[\!\left[\frac{v_i''}{v_{i+1}} \right]\!\right] = \frac{1}{2} \left[\!\left[\frac{v_i}{v_{i+1}} \frac{v_{i-1}}{v_i} \frac{v_i''}{v_{i-1}} \right]\!\right] = \left[\!\left[\frac{b_i b_{i-1}}{\bar{b}_{i-1}} \right]\!\right] = b_i + b_{i-1} - \bar{b}_{i-1}.$$

Using the previous result (A.10) for $\bar{b}_{i-1}$, we get

$$\bar{b}_i = b_i - (a_i - a_{i-1})^+ + \left(((a_i - a_{i-1})^+ - b_i)^+ - (a_i - a_{i-1}) + ((a_i - a_{i-1})^+ + b_{i-1})^+ \right)^+$$

or in max-plus notation

$$\bar{b}_i = \left[\!\left[\frac{1 + \frac{a_{i-1}}{a_i} \left(1 + \frac{1}{b_i} \left(1 + \frac{a_i}{a_{i-1}} \right) \right) \left(1 + b_{i-1} \left(1 + \frac{a_i}{a_{i-1}} \right) \right)}{\frac{1}{b_i} \left(1 + \frac{a_i}{a_{i-1}} \right)} \right]\!\right]. \quad (A.11)$$

At this stage, we have generated the update rules for Dynnikov coordinates for a counterclockwise (CCW) exchange of points (corresponding to the σ_i^{-1} generator). Here are all the rules together:

$$\bar{a}_{i-1} = a_{i-1} + s_1, \qquad\qquad \bar{a}_i = a_i - s_2,$$
$$\bar{b}_{i-1} = b_{i-1} + t, \qquad\qquad \bar{b}_i = b_i - t,$$

with the auxiliary variables

$$r = (a_i - a_{i-1}), \qquad\qquad t = r^+ - (s_1 - r + s_2)^+,$$
$$s_1 = (r^+ + b_{i-1})^+, \qquad\qquad s_2 = (r^+ - b_i)^+.$$

This is a superficially different (though ultimately equivalent) set of update rules from those given by Eqs. (A.2)–(A.3). These appear a bit more symmetric; note that each updated variable is given as the old variable plus some quantity.

A.3.2 Equivalence of Update Rules

The update rules that we have derived are superficially different from the ones given in Sect. 8.2. Here we show directly the equivalence of the update rules.

First for $\bar{a}_{i-1}$, we write (8.19a) in max-plus notation:

$$\left[\!\left[a_{i-1}(1 + b_{i-1}) \left(1 + \frac{a_i(1 + b_i)}{a_{i-1}(1 + 1/b_{i-1})(1 + b_i)} \right) \right]\!\right] = \left[\!\left[a_{i-1} \left(1 + b_{i-1} + \frac{a_i b_{i-1}}{a_{i-1}} \right) \right]\!\right]$$

which is equal to (A.7).

Next for $\bar{a}_i$, we write (8.19c) in max-plus notation:

$$
\left[\!\!\left[\frac{a_i}{(1+1/b_i)\left(1+\frac{(1+1/b_{i-1})a_i}{a_{i-1}(1+1/b_{i-1})(1+b_i)}\right)}\right]\!\!\right] = \left[\!\!\left[\frac{a_i}{\left(1+1/b_i+\frac{a_i}{a_{i-1}b_i}\right)}\right]\!\!\right]
$$

$$
= \left[\!\!\left[\left(\frac{1}{a_i}\left(1+\left(1+\frac{a_i}{a_{i-1}}\right)\frac{1}{b_i}\right)\right)^{-1}\right]\!\!\right],
$$

which is equal to (A.8).

For $\bar{b}_{i-1}$, we write (8.19b) in max-plus notation and try to equate to (A.9):

$$
\left[\!\!\left[\frac{b_i}{1+\frac{a_{i-1}}{a_i}(1+b_i)(1+1/b_{i-1})}\right]\!\!\right] =
$$

$$
\left[\!\!\left[\left(\frac{\left(1+\frac{a_{i-1}}{a_i}\left(1+\frac{1}{b_i}\left(1+\frac{a_i}{a_{i-1}}\right)\right)\right)\left(1+b_{i-1}\left(1+\frac{a_i}{a_{i-1}}\right)\right)}{b_{i-1}\left(1+\frac{a_i}{a_{i-1}}\right)}\right)^{-1}\right]\!\!\right].
$$

Invert both sides:

$$
\left[\!\!\left[\frac{1}{b_i}\left(1+\frac{a_{i-1}}{a_i}(1+b_i)(1+1/b_{i-1})\right)\right]\!\!\right] = \left[\!\!\left[\frac{\frac{a_{i-1}}{a_i}+\frac{a_{i-1}}{a_i}\left(\frac{1}{b_i}+b_{i-1}+\frac{b_{i-1}}{b_i}\left(1+\frac{a_i}{a_{i-1}}\right)\right)}{b_{i-1}}\right]\!\!\right]
$$

$$
= \left[\!\!\left[\frac{\frac{a_{i-1}b_i}{a_ib_{i-1}}+\frac{a_{i-1}}{a_ib_{i-1}}+\frac{a_{i-1}b_i}{a_i}+\frac{a_{i-1}}{a_i}+1}{b_i}\right]\!\!\right]
$$

$$
= \left[\!\!\left[\frac{1+\frac{a_{i-1}}{a_i}\left(\frac{b_i}{b_{i-1}}+\frac{1}{b_{i-1}}+b_i+1\right)}{b_i}\right]\!\!\right]
$$

which are indeed equal.

For $\bar{b}_i$, we need not compare. Both the Dynnikov update rules from 8.2 and in this appendix satisfy $\bar{b}_i+\bar{b}_{i-1}=b_i+b_{i-1}$, so checking equivalence for $\bar{b}_{i-1}$ is sufficient.

A.3.3 Clockwise Switch

Fortunately, we do not need to repeat this analysis for the case of a clockwise switch (σ_i). The Dynnikov coordinate update rules can be derived from the CCW case using conjugation by a vertical mirror symmetry. Consider the mirror symmetry operator that switches μ_{2i} and μ_{2i-1} for all i. This is equivalent to replacing a_i with $-a_i$. If we apply this operator, use the CCW update rules, and then apply the vertical mirror symmetry operator again, we get

$$\bar{a}_{i-1} = a_{i-1} - y_1, \qquad \bar{a}_i = a_i + y_2, \qquad \bar{b}_{i-1} = b_{i-1} + z, \qquad \bar{b}_i = b_i - z$$

with the auxiliary variables

$$x = (a_{i-1} - a_i), \qquad\qquad z = x^+ - (y_1 - r + y_2)^+,$$
$$y_1 = (x^+ + b_{i-1})^+, \qquad\qquad y_2 = (x^+ - b_i)^+.$$

Again, these rules are equivalent to those previously given, Eqs. (8.12)–(8.13).

A.3.4 Edge Cases

The CW and CCW cases for switches of the first and last pair of points also follow from what we have already shown. We can add some extra curves to the Dynnikov disk at the edges, which allow us to directly apply the update rules we have previously derived. There are some added relations that hold (as part of adding the extra curves), which simplify the resultant update rules considerably.

References

1. Adler RL, Konheim AG, McAndrew MH (1965) Topological entropy. Trans Am Math Soc 114(2):309–319
2. Akpulat M, Ekinci M (2019) Detecting interaction/complexity within crowd movements using braid entropy. Front Inf Technol Electron Eng 20(6):849–861
3. Allshouse MR, Thiffeault JL (2012) Detecting coherent structures using braids. Physica D 241(2):95–105
4. Aref H (1984) Stirring by chaotic advection. J Fluid Mech 143:1–21
5. Aref H, Blake JR, Budišić M, Cardoso SS, Cartwright JH, Clercx HJ, El Omari K, Feudel U, Golestanian R, Gouillart E, van Heijst GF, Krasnopolskaya TS, Le Guer Y, MacKay RS, Meleshko VV, Metcalfe G, Mezić I, de Moura AP, Piro O, Speetjens MFM, Sturman R, Thiffeault JL, Tuval I (2017) Frontiers of chaotic advection. Rev Mod Phys 89(2):025007
6. Arnold VI, Avez A (1968) Ergodic problems of classical mechanics. W. A. Benjamin, New York
7. Artin E (1925) Theorie der zöpfe. Abh Math Semin Univ Hambg 4(1):47–72
8. Artin E (1947) Theory of braids. Ann Math 48(1):101–126
9. Band G, Boyland PL (2007) The Burau estimate for the entropy of a braid. Algeb Geom Topol 7:1345–1378
10. Bangert PD, Berger MA, Prandi R (2002) In search of minimal random braid configurations. J Phys A 35(1):43–59
11. Bell M, Schleimer S (2017) Slow north-south dynamics on $\mathscr{PML}$. Groups Geom Dynam 11(3):1103–1112
12. Bestvina M, Handel M (1995) Train-tracks for surface homeomorphisms. Topology 34(1):109–140
13. Bigelow SJ (1999) The Burau representation is not faithful for $n = 5$. Geom Topol 3:397–404
14. Bigelow SJ (2001) Braid groups are linear. J Am Math Soc 14(2):471–486
15. Binder BJ (2010) Ghost rods adopting the role of withdrawn baffles in batch mixer designs. Phys Lett A 374:3483–3486

16. Binder BJ, Cox SM (2008) A mixer design for the pigtail braid. Fluid Dyn Res 40:34–44

17. Birman J, Brinkmann P, Kawamuro K (2012) Polynomial invariants of pseudo-Anosov maps. J Topo Anal 4(1):13–47

18. Birman JS (1975) Braids, links and mapping class groups. Annals of mathematics studies, vol 82. Princeton University Press, Princeton, NJ

19. Birman JS, Brendle TE (2005) Braids: A survey. In: Menasco W, Thistlethwaite M (eds) Handbook of Knot Theory. Elsevier, Amsterdam, pp 19–104

20. Biryukov ON (2007) A bound for the topological entropy of homeomorphisms of a punctured two-dimensional disk. J Math Sci 146(1):5483–5489, translated from Fundamentalnaya i Prikladnaya Matematika, vol 11(5), pp 47–55, 2005

21. Bowen R (1971) Entropy for group endomorphisms and homogeneous spaces. Trans Am Math Soc 153:401–414

22. Bowen R (1971) Periodic points and measures for axiom A diffeomorphisms. Trans Am Math Soc 153:377–397

23. Bowen R (1978) Entropy and the fundamental group. In: Structure of Attractors. Lecture notes in mathematics, vol 668. Springer, New York, pp 21–29

24. Boyland PL (1994) Topological methods in surface dynamics. Topol Appl 58:223–298

25. Boyland PL, Franks JM (1989) Lectures on dynamics of surface homeomorphisms. University of Warwick Preprint, notes by C. Carroll, J. Guaschi and T. Hall

26. Boyland PL, Harrington J (2011) The entropy efficiency of point-push mapping classes on the punctured disk. Algeb Geom Topol 11(4):2265–2296

27. Boyland PL, Aref H, Stremler MA (2000) Topological fluid mechanics of stirring. J Fluid Mech 403:277–304

28. Boyland PL, Stremler MA, Aref H (2003) Topological fluid mechanics of point vortex motions. Physica D 175:69–95

29. Budišić M, Thiffeault JL (2015) Finite-time braiding exponents. Chaos 25:087407

30. Burau W (1936) Über Zopfgruppen und gleichsinnig verdrilte Verkettungen. Abh Math Semin Univ Hambg 11:171–178

31. Butkovič P (2010) Max-linear systems: theory and algorithms. Springer, London

32. Candelaresi S, Pontin DI, Hornig G (2017) Quantifying the tangling of trajectories using the topological entropy. Chaos 27(9):093102

33. Casson AJ, Bleiler SA (1988) Automorphisms of surfaces after Nielsen and Thurston. London mathematical society student texts, vol 9. Cambridge University Press, Cambridge

34. Caussin JB, Bartolo D (2015) Braiding a flock: winding statistics of interacting flying spins. Phys Rev Lett 114(25):258101

35. Cha JC (2011) CBraid: A C++ library for computations in braid groups. https://github.com/jeanluct/cbraid

36. Childress S, Gilbert AD (1995) Stretch, twist, fold: the fast dynamo. Springer, Berlin

37. Connelly RK, Valenti-Jordan J (2008) Mixing analysis of a Newtonian fluid in a 3D planetary pin mixer. Chem Eng Res Design 86(12):1434–1440
38. Dehornoy P (2008) Efficient solutions to the braid isotopy problem. Discr Appl Math 156:3091–3112
39. Di Labbio G, Thiffeault JL, Kadem L (2022) Braids in the heart: Global measures of mixing for cardiovascular flows. Preprint
40. Dickinson HM (1906) Candy-pulling machine. Cooperative Classification A23G3/10
41. Dinaburg EI (1970) The relation between topological and metric entropy. Soviet Math Dokl 11:13–16
42. Dinaburg EI (1971) On the relations among various entropy characteristics of dynamical systems. Math USSR Izvestija 5(2):337–378
43. Dynnikov IA (2002) On a Yang–Baxter map and the Dehornoy ordering. Russ Math Surv 57(3):592–594
44. Dynnikov IA, Wiest B (2007) On the complexity of braids. J Eur Math Soc 9(4):801–840
45. Farb B, Margalit D (2011) A primer on mapping class groups. Princeton University Press, Princeton, NJ
46. Fathi A, Laundenbach F, Poénaru V (1979) Travaux de Thurston sur les surfaces. Astérisque 66-67:1–284
47. Fathi A, Laundenbach F, Poénaru V (2012) Thurston's work on surfaces. Princeton University Press, Princeton, NJ, translation of the 1979 original by D. M. Kim and D. Margalit
48. Filippi M, Budišić M, Allshouse MR, Atis S, Thiffeault JL, Peacock T (2020) Using braids to quantify interface growth and coherence in a rotor-oscillator flow. Phys Rev Fluids 5:054504
49. Finn MD, Thiffeault JL (2007) Topological entropy of braids on the torus. SIAM J Appl Dyn Sys 6:79–98
50. Finn MD, Thiffeault JL (2011) Topological optimization of rod-stirring devices. SIAM Rev 53(4):723–743
51. Finn MD, Cox SM, Byrne HM (2003) Chaotic advection in a braided pipe mixer. Phys Fluids 15(11):L77–L80
52. Finn MD, Cox SM, Byrne HM (2003) Topological chaos in inviscid and viscous mixers. J Fluid Mech 493:345–361
53. Finn MD, Thiffeault JL, Gouillart E (2006) Topological chaos in spatially periodic mixers. Physica D 221(1):92–100
54. Firchau PJG (1893) Machine for working candy. Cooperative Classification B01F7/166
55. Fox RH (1953) Free differential calculus. I: derivation in the free group ring. Ann Math 57(3):547–560
56. Francois N, Xia H, Punzmann H, Faber B, Shats M (2015) Braid entropy of two-dimensional turbulence. Sci Rep 5:18564
57. Fried D (1986) Entropy and twisted cohomology. Topology 25(4):455–470
58. Funar L, Kohno T (2013) On Burau's representations at roots of unity. Geom Ded 169(1):145–163

59. Gambaudo JM, Ghys E (2005) Braids and signatures. Bulletin de la Société mathématique de France 133(4):541–579

60. Gouillart E, Finn MD, Thiffeault JL (2006) Topological mixing with ghost rods. Phys Rev E 73:036311

61. Gouillart E, Kuncio N, Dauchot O, Dubrulle B, Roux S, Thiffeault JL (2007) Walls inhibit chaotic mixing. Phys Rev Lett 99:114501

62. Gouillart E, Dauchot O, Dubrulle B, Roux S, Thiffeault JL (2008) Slow decay of concentration variance due to no-slip walls in chaotic mixing. Phys Rev E 78:026211

63. Gouillart E, Dauchot O, Thiffeault JL, Roux S (2009) Open-flow mixing: experimental evidence for strange eigenmodes. Phys Fluids 21(2):022603

64. Hall T (2012) Train: A C++ program for computing train tracks of surface homeomorphisms. http://www.liv.ac.uk/~tobyhall/T_Hall.html

65. Hall T (2013) The Bestvina–Handel algorithm, unpublished

66. Hall T, Yurttaş SÖ (2009) On the topological entropy of families of braids. Topol Appl 156(8):1554–1564

67. Handel M (1985) Global shadowing of pseudo-Anosov homeomorphisms. Ergod Th Dynam Sys 8:373–377

68. Hansen VL (1993) Jakob Nielsen (1890–1959). Math Intell 15(4):44–53

69. Hatcher A (2001) Algebraic topology. Cambridge University Press, Cambridge

70. Haynes PH, Vanneste J (2005) What controls the decay of passive scalars in smooth flows? Phys Fluids 17:097103

71. Heidergott B, Olsder GJ, van der Woude JW (2006) Max Plus at work: modeling and analysis of synchronized systems: a course on Max-Plus algebra and its applications. Princeton University Press, Princeton, NJ

72. Kassel C, Turaev V (2008) Braid groups. Springer, New York, NY

73. Kirsch E (1928) Candy-pulling machine. U.S. Classification 366/70; International Classification A23G3/10, A23G3/02; Cooperative Classification A23G3/10; European Classification A23G3/10

74. Kobayashi T, Umeda S (2010) A design for pseudo-Anosov braids using hypotrochoid curves. Topol Appl 157:280–289

75. Kolev B (1989) Entropie topologique et représentation de Burau. C R Acad Sci Sér I 309(13):835–838, English translation at http://arxiv.org/abs/math.DS/0304105.

76. Lanneau E, Thiffeault JL (2011) On the minimum dilatation of braids on the punctured disc. Geom Ded 152(1):165–182

77. Lanneau E, Thiffeault JL (2011) On the minimum dilatation of pseudo-Anosov homeomorphisms on surfaces of small genus. Ann Inst Four 61(1):105–144

78. Lawrence RJ (1990) Homological representations of the Hecke algebra. Commun Math Phys 135(1):141–191

79. Long DD (1986) A note on the normal subgroups of mapping class groups. Math Proc Cambridge Phil Soc 99(1):79–87

80. Long DD, Paton M (1993) The Burau representation is not faithful for $n \geq 6$. Topology 32:439–447

81. Lynch JC (2019) Pseudo-Anosov homeomorphisms with large topological entropy. PhD thesis, University of Wisconsin – Madison
82. Manning A (1978) Topological entropy and the first homology group. In: Dynamical systems — Warwick 1974. Lecture notes in mathematics, vol 468. Springer, Berlin, pp 185–190
83. Masur H, Smillie J (1993) Quadratic differentials with prescribed singularities and pseudo-Anosov diffeomorphisms. Comment Math Helv 68(2):289–307
84. Mavrogiannis C, Knepper RA (2019) Multi-agent path topology in support of socially competent navigation planning. Int J Robot Res 38(2-3):338–356
85. McEneaney WM (2006) Max-plus methods for nonlinear control and estimation. Birkhauser, Boston, MA
86. Moody JA (1991) The Burau representation of the braid group B_n is unfaithful for large n. Bull Am Math Soc 25:379–384
87. Mosher L (2003) What is a train track? Math Intell 50(3):354–356
88. Moussafir JO (2006) On computing the entropy of braids. Func Anal and Other Math 1(1):37–46
89. Munkres JR (2000) Topology, 2nd edn. Prentice-Hall, Upper Saddle River, NJ
90. Newhouse SE (1988) Entropy and volume. Ergod Th Dynam Sys 8:283–299
91. Newhouse SE (1991) Entropy in smooth dynamical systems. In: Proceedings of the international congress of mathematicians, vol. I, II (Kyoto, 1990). Math. Soc. Japan, Tokyo, pp 1285–1294
92. Newhouse SE, Pignataro T (1993) On the estimation of topological entropy. J Stat Phys 72(5-6):1331–1351
93. Nielsen J (1944) Surface transformation classes of algebraically finite type. Danske Vid. Selsk. Math.-Phys. Medd., vol 21(2). Munksgaard
94. Nitz CGW (1918) Candy-puller. U.S. Classification 366/70; Cooperative Classification A23G3/10
95. Oseledec VI (1968) A multiplicative theorem: Lyapunov characteristic numbers for dynamical systems. Trans Moscow Math Soc 19:197–231
96. Ottino JM (1989) The kinematics of mixing: stretching, chaos, and transport. Cambridge University Press, Cambridge
97. Paterson MS, Razborov AA (1991) The set of minimal braids is co-NP complete. J Algorithm 12:393–408
98. Penner RC, Harer JL (1991) Combinatorics of train tracks. Annals of mathematics studies, vol 125. Princeton University Press, Princeton, NJ
99. Pierrehumbert RT (1994) Tracer microstructure in the large-eddy dominated regime. Chaos Solitons Fract 4(6):1091–1110
100. Puckett JG, Lechenault F, Daniels KE, Thiffeault JL (2012) Trajectory entanglement in dense granular materials. J Stat Mech Theory Exp 2012(6):P06008
101. Ramsey A, Furey H (2004) WOCE subsurface float data assembly center. https://www.aoml.noaa.gov/phod/float_traj/
102. Richards FH (1905) Process of making candy. Cooperative Classification A23G3/42
103. Roberts E, Sindi S, Smith SA, Mitchell KA (2019) Ensemble-based topological entropy calculation (E-tec). Chaos 29(1):013124

104. Rolfsen D (2003) New developments in the theory of Artin's braid groups. Topol Appl 127(1-2):77–90

105. Rothstein D, Henry E, Gollub JP (1999) Persistent patterns in transient chaotic fluid mixing. Nature 401(6755):770–772

106. Seneta E (1981) Non-negative Matrices and Markov Chains. Springer, New York

107. Shendruk TN, Doostmohammadi A, Thijssen K, Yeomans JM (2017) Dancing disclinations in confined active nematics. Soft Matter 13(21):3853–3862

108. Smith SA, Dunn S (2021) Topological entropy of surface braids and maximally efficient mixing. SIAM J Appl Dyn Sys. Preprint

109. Smith SA, Warrier S (2016) Reynolds number effects on mixing due to topological chaos. Chaos 26:033106

110. Steenrod NE, Halmos PH, Schiffer MM, Dieudonné JA (1973) How to write mathematics. American Mathematical Society, Providence, RI

111. Stremler MA, Chen J (2007) Generating topological chaos in lid-driven cavity flow. Phys Fluids 19:103602

112. Sturman R, Ottino JM, Wiggins S (2006) The mathematical foundations of mixing: the linked twist map as a paradigm in applications: micro to macro, fluids to solids. Cambridge University Press, Cambridge

113. Sukhatme J, Pierrehumbert RT (2002) Decay of passive scalars under the action of single scale smooth velocity fields in bounded two-dimensional domains: from non-self-similar probability distribution functions to self-similar eigenmodes. Phys Rev E 66:056302

114. Tan AJ, Roberts E, Smith SA, Olvera UA, Arteaga J, Fortini S, Mitchell KA, Hirst LS (2019) Topological chaos in active nematics. Nat Phys 15(10):1033–1039

115. Taylor CK, Llewellyn Smith SG (2016) Dynamics and transport properties of three surface quasigeostrophic point vortices. Chaos 26(11):113117

116. Thiffeault JL (2005) Measuring topological chaos. Phys Rev Lett 94(8):084502

117. Thiffeault JL (2010) Braids of entangled particle trajectories. Chaos 20:017516

118. Thiffeault JL (2018) The mathematics of taffy pullers. Math Intell 40(1):26–35

119. Thiffeault JL, Budišić M (2014) Braidlab: A software package for braids and loops. http://arxiv.org/abs/1410.0849

120. Thiffeault JL, Childress S (2003) Chaotic mixing in a torus map. Chaos 13(2):502–507

121. Thiffeault JL, Finn MD (2006) Topology, braids, and mixing in fluids. Philos Trans R Soc Lond A 364:3251–3266

122. Thiffeault JL, Finn MD, Gouillart E, Hall T (2008) Topology of chaotic mixing patterns. Chaos 18:033123

123. Thiffeault JL, Gouillart E, Dauchot O (2011) Moving walls accelerate mixing. Phys Rev E 84(3):036313

124. Thurston D (2012) The geometry and algebra of curves on surfaces. https://math.berkeley.edu/~qchu/Notes/274/. Accessed 2016-01-08

125. Thurston WP (1988) On the geometry and dynamics of diffeomorphisms of surfaces. Bull Am Math Soc 19:417–431. Circulated as a preprint in 1976
126. Thurston WP (2002) The geometry and topology of three-manifolds. Lecture notes, version 1.1. http://library.msri.org/books/gt3m/; original notes from 1980
127. Tumasz SE (2012) Topological stirring. PhD thesis, University of Wisconsin – Madison, Madison, WI
128. Tumasz SE, Thiffeault JL (2013) Estimating topological entropy from the motion of stirring rods. Procedia IUTAM 7:117–126
129. Tumasz SE, Thiffeault JL (2013) Topological entropy and secondary folding. J Nonlinear Sci 13(3):511–524
130. Vanneste J (2006) Intermittency of passive-scalar decay: Strange eigenmodes in random shear flows. Phys Fluids 18:087108
131. Venkataramana TN (2014) Image of the Burau representation at d-th roots of unity. Ann Math 179(3):1041–1083
132. Yeung M, Cohen-Steiner D, Desbrun M (2020) Material coherence from trajectories via Burau eigenanalysis of braids. Chaos 30:033122
133. Yomdin Y (1987) Volume growth and entropy. Israel J Math 57(3):285–300

Index

© The Author(s), under exclusive license to Springer Nature Switzerland AG 2022
J.-L. Thiffeault, *Braids and Dynamics*, Frontiers in Applied Dynamical Systems:
Reviews and Tutorials 9, https://doi.org/10.1007/978-3-031-04790-9

Printed in the United States
by Baker & Taylor Publisher Services